W9-DEF-478

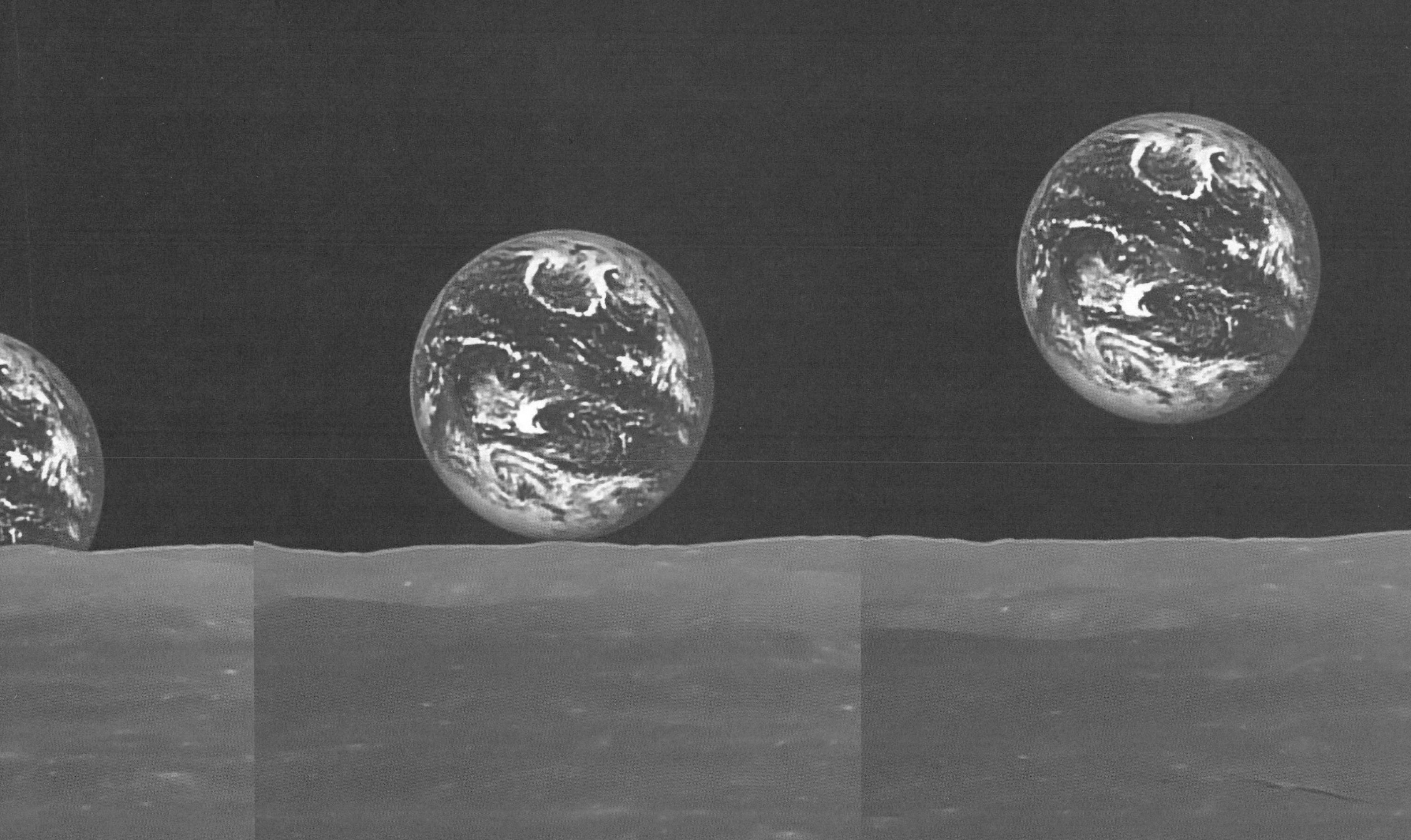

MARKETING
THE MOON

MARKETING

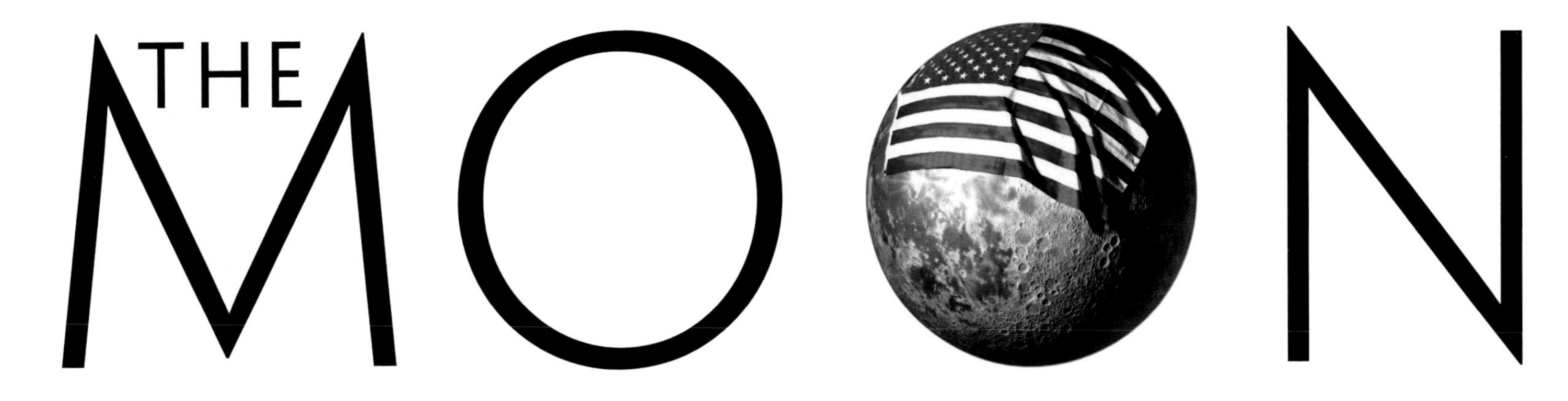

THE MOON

THE SELLING of the APOLLO LUNAR PROGRAM

DAVID MEERMAN SCOTT and **RICHARD JUREK**

with a foreword by **CAPTAIN EUGENE A. CERNAN**

THE MIT PRESS | CAMBRIDGE, MASSACHUSETTS AND LONDON, ENGLAND

MIT Press books may be purchased at special quantity discounts for business or sales promotional use. For information, please email: special_sales@mitpress.mit.edu.

LIBRARY OF CONGRESS CATALOGING-IN-PUBLICATION DATA
Scott, David Meerman.
Marketing the moon : the selling of the Apollo lunar program / David Meerman Scott and Richard Jurek.
pages cm
Includes bibliographical references and index.
ISBN 978-0-262-02696-3 (hardcover : alk. paper)
1. Project Apollo (U.S.)—Public relations—History. 2. United States. National Aeronautics and Space Administration—Public relations. 3. Space flight to the moon—History. 4. Moon—Exploration—History. 5. Astronautics—Press coverage—United States—History.
I. Jurek, Richard. II. Title.
TL789.8.U6A58156 2014
659.2'9629454—dc23

2013039798

Produced, designed, composed, and edited by Scott-Martin Kosofsky at The Philidor Company, Lexington, Massachusetts. www.philidor.com
Chief editor: Alan Andres.

Printed and bound in China by Everbest.
10 9 8 7 6 5 4 3 2 1

Contents

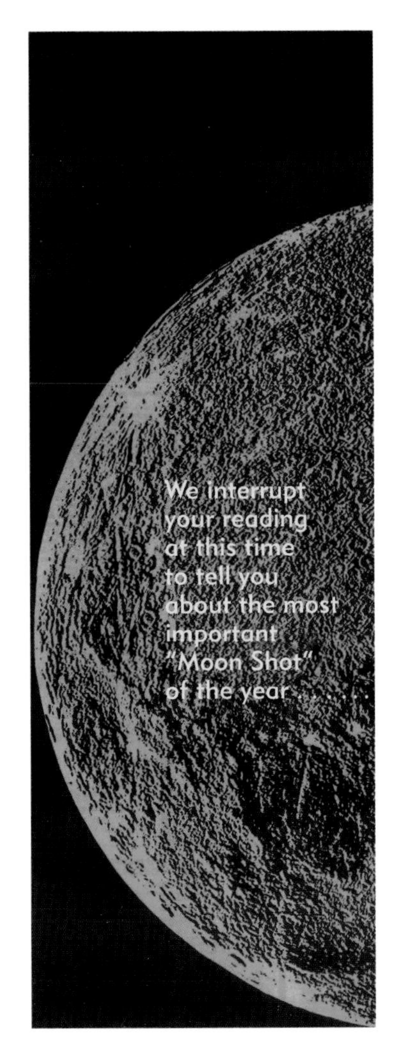
We interrupt your reading at this time to tell you about the most important "Moon Shot" of the year.......

Foreword

"We were . . . marketing the United States of America"

By Captain Eugene A. Cernan, USN (*ret.*)

THE GOAL of the Apollo program was to send a man to the Moon and return him safely to Earth, and do it before the end of the 1960s. Without fully realizing it at the time, we took those first steps to the Moon with the first manned mission of Project Mercury, when Alan Shepard flew his sub-orbital flight on May 5, 1961, becoming the first American in space. It was but three weeks later, on May 25, 1961, that President John F. Kennedy stood before Congress and said, "First, I believe that this nation should commit itself to achieving the goal, before this decade is out, of landing a man on the Moon and returning him safely to the Earth. No single space project in this period will be more impressive to mankind, or more important for the long-range exploration of space; and none will be so difficult or expensive to accomplish." That sealed what became our goal and commitment leading to the historic landing of Apollo 11.

Along the way, and totally unexpected by us, we astronauts became very visible public figures. This wasn't NASA's initial intent, but they adapted quickly. It was the press, and in turn the public, who declared us "heroes," and from that followed the inevitable responsibility to "market" the space program, both to Congress and to the public that elected it. Even before many of us had flown into space, before we had done anything to deserve such adulation, we became icons of our time. We astronauts, however, were just the tip of the arrow; behind us were hundreds of thousands of men and women—the real heroes of the era—who were the strength behind the bow. Yet, from the moment of liftoff until the time we returned to Earth, the astronauts were, and continue to be, the face of America's space program. It was, in part, a marketing decision that gave the world access through reports in the press and the live audio and television feeds from Houston, the Cape, and from space itself. During the high point of Apollo, the public just couldn't get enough.

Because of the public's desire, we were called upon to share with the nation why we felt what we were doing had value; in fact, people began to demand no less. As a consequence, NASA would send us traveling throughout the country, one at a time, on what was called our "week in the barrel," public affairs duty, which allowed the others to concentrate on their mission planning and training. In addition, we went on goodwill tours after each of our missions, not only to major cities in the United States, but to world capitals as well.

Thus, the astronauts became the leading edge of one of the biggest marketing efforts in history. We were the voices of a nation marketing the United States of America. We were whatever the people imagined us to be. There were other marketing efforts by NASA and the aerospace contractors, but no matter how intense their efforts were, the astronauts were the ones people wanted to hear and see. A personal association with someone who "had been there" is what everyone wanted. Even now we are asked the same questions we were asked decades ago: "What was it like?" "What did the Earth look like from space?" "Did you feel close to God?" I believed then—and I believe now—that it is our responsibility to give back to the nation that gave us an opportunity to go where no man had gone before. It may be our legacy.

Looking back, I must admit I learned something from those weeks "in the barrel." If your desire is to bring others into your camp, they must know that you, yourself, believe in what they are hearing; your sincerity and passion must be evident. You must share your ideas with them—not talk at them—if you want to achieve your ultimate goal. To me, *that* is marketing and it was a key part of the U.S. space program that garnered the support of the American people. Unfortunately, as a casualty of this vast success, the support began to fade. What you're about to read, *Marketing the Moon*, is a story of the challenges and ultimate success of marketing one of the greatest achievements in American—and world—history. ⊙

During twenty years as a Naval Aviator, including thirteen years with NASA, Captain Eugene A. Cernan left his mark on history with three historic missions in space: as the pilot of Gemini 9, the lunar module pilot of Apollo 10, and the commander of Apollo 17. Flying to the Moon not once, but twice, he also holds the distinction of being the second American to walk in space and the last man to have left his footprints on the lunar surface.

Introduction

"Because without public relations . . . we would have been unable to do it."

—Wernher von Braun, July 22, 1969

THROUGHOUT THE LAUNCH and the long flight of Apollo 11 to the Moon, Dr. Wernher von Braun, director of NASA's Marshall Spaceflight Center and one of the greatest media stars of the space age, remained uncharacteristically silent. The chief architect of the Apollo Saturn V launch rocket did not participate in briefings and interviews after the successful landing of the lunar module *Eagle*. Neither was there a public word from him after Neil Armstrong and Buzz Aldrin made their historic moonwalk, nor after *Eagle*'s liftoff from the lunar surface and rendezvous with Michael Collins in the command module *Columbia*. Von Braun chose to speak to the press only after *Columbia* was well on its way back to Earth.

Finally, at 12:34 P.M., Houston time, on July 22, the world's most renowned rocket scientist addressed the world's journalists. "I would like to thank all of you for all of the fine support you have always given to the program," von Braun said from the podium at NASA's Manned Spacecraft Center, packed with reporters and camera crews. "Because without public relations and good presentations of these programs to the public, we would have been unable to do it."[1]

When America made real John F. Kennedy's goal of becoming the first nation to place a man on the Moon before the end of the 1960s, the chief technological architect of that success chose not to address the public directly, but to first thank the assembled press. What von Braun acknowledged in that moment was that, without strong public support abetted by high-profile, enthusiastic public relations, the entire achievement and its funding would have been unthinkable. Von Braun was well aware that possible failure and catastrophe lurked a step away from every success, and that to overcome it would require the forbearance of the press, as congressional and public support might not exist without it. He was hardly the only program official to recognize this reality and hedge his bets accordingly. Between the years 1963 and 1969, America spent an average of 3.3% of its budget on NASA; from 1965 to 1968, it surpassed 4%. And it was likely more, as Department of Defense expenditures on the program were not part of the official budget line. (The spacecraft recovery efforts alone for each mission involved ships, aircraft, and thousands of Navy and other DoD personnel.) When reservations over the costs were raised by a number of legislators, President Lyndon B. Johnson was quick to reply, "Now, would you rather have us be a second-rate nation or should we spend a little money?"[2] The vast space program expenditures coincided with another costly—and far more controversial—government endeavor, the Vietnam War, and it also coincided with the commencement of the Great Society social programs. Though not without its detractors, the space program enjoyed the highest public approval throughout the 1960s.

APOLLO IS THE LARGEST, and we believe the most important, marketing and public relations case study in history. It's a story that needed to be told but to date had not.

The concepts of marketing and public relations are often used interchangeably, even by those who are involved in the field. There are many definitions of both terms, but simply, "marketing" is a multidisciplinary process by which a company or institution actively promotes, sells, or distributes a product, idea, or service to potential customers. "Public relations," on the other hand, is a process (an aspect of marketing, in fact), by which a company or an institution tries to encourage broad, public understanding and acceptance of an idea, product, or service among its various potential audiences. In the Apollo program, the marketing was most often handled by the contractors and subcontractors, as they had reason not only to get out the story of their involvement in the program, but also to sell their capabilities on both the national and international stages. The PR responsibilties, while also important to the contractors, remained primarily with the NASA Public Affairs Office, as it was incumbent on them to sustain public and congressional interest in the program. Viewed together, the marketing and PR of Apollo represent a singular and eminently instructive case study for modern-day practitioners.

We are both professional marketers who are also space enthusiasts and collectors of Apollo lunar program material, including items flown to the Moon and used on the missions.

OPPOSITE: Tickertape parade in New York City, August 13, 1969, following the Apollo 11 mission to the Moon. In the backseat of the first car (a Chrysler Imperial Parade Phaeton, one of only three made) are Neil Armstrong, Edwin "Buzz" Aldrin, and Michael Collins.

1. Wernher von Braun quoted by Charles K. Siner, Jr., "You and Neil and Buzz—You Made It" in *The Quill*, September 1969.
2. Edward C. Welch, Oral History, pp. 11–12, Lyndon B. Johnson Presidential Library, Austin, Texas.

Over the years we have independently researched the marketing and public relations aspects of the Apollo program and carefully acquired items into our collections, which include hundreds of contractor press materials and NASA documents. We met several times a year at Apollo astronaut gatherings, talking about the intersection of our interests in space and marketing. It was at the 40th anniversary dinner celebrating the Apollo 13 mission, in 2010, with many Apollo astronauts in attendance, including Fred Haise and Jim Lovell of Apollo 13, that the idea for this book was hatched. We knew we were onto something when we shared our ideas with astronauts, NASA officials, contractor employees, and members of the media from the era, all of whom were unabashedly enthusiastic and forthcoming. At a later event, we ran it by Buzz Aldrin, who was wholeheartedly supportive, proceeding immediately to share stories and thoughts on the marketing of space, past, present, and future. He spoke with particular passion about the benefits derived by NASA and the nation from the Apollo 11 public relations world tour that followed his mission. At that moment we knew we had a story that needed to be told.

Our goal for *Marketing the Moon* has been to examine the inner workings and public perceptions of the Apollo lunar program through the lens of practicing PR and marketing professionals. We do not attempt an encyclopedic presentation, but rather an analysis of what was done, and what worked and what did not. We have been driven to the Apollo program not only because of its inherent historical significance, but by the highly unsual nature of how it unfolded through the unprecedented cooperation and teamwork of government, industry, media, and over 400,000 people working together toward the achievement of a common goal. We were drawn to the compelling and sometimes unexpected (and even counterintuitive) stories we heard from people who worked behind the scenes in this often overlooked aspect of the program.

We also wrote *Marketing the Moon* because the critical public relations and marketing of the program have often been mischaracterized. We chose to write it now because the story was not, by and large, part of official government records, and much of the salient information could be gathered only while many of the key participants were still alive and willing to be interviewed and share their personal stories and archives. Much of the material we gathered was acquired directly from the participants; other pieces were purchased at auction, and still others bought or traded within the small, but dedicated, group of collectors, many of whom are members of the collectSPACE online community. Some of our collections are open to the public via the Internet so others can learn from them.[3]

The materials used in the marketing of Apollo were wide and deep, as hundreds of corporations, academic institutions, and government agencies were involved in highly technical work, which required the promulgation of detailed knowledge and data to an extent that changed forever the roles of journalists and public relations professionals. Terminology, some of it entirely new, and professional roles that were hitherto the exclusive realms of scientists, engineers, and manufacturers, were suddenly part of the public discourse. Thousands of photographs, diagrams and drawings, hundreds of hours of film and television, countless in-person press briefings, and press kits that could run to more than a hundred pages, provided reporters an unprecedented luxury of information. As a result, science and technology reporting became an essential part of journalism. "Covering the space program presented a challenge to us all," said Walter Cronkite, managing editor of *CBS Evening News*. "There was a great deal we had to learn about the mechanics of space flight and the idiosyncrasies of the physics of moving bodies in the weightlessness and atmosphere-free environment of space."[4] Every need for information and explanation was anticipated with profuse and well-prepared materials. How this was done and by whom are among this book's key themes.

Some writers and historians have described how NASA, itself, was a big PR machine that developed the grand marketing scheme, choreographing its every detail. We will show that was not at all the case. James Kauffman's *Selling Outer Space*[5] and Megan Garber, writing in *The Atlantic*,[6] to name just two, focus relentlessly on the *Life* magazine contracts with the astronauts, but fail to explain NASA's benign, if naive, motivation for permitting them. While the stories that appeared in *Life* make for a convenient focal point, and are illustrative on a certain level, a more balanced view reveals that no one outlet "controlled" the image of the astronauts.

NASA did not have a massive PR machine that worked to shape the global press image of the astronauts and the pro-

3. www.apolloartifacts.com, www.jeffersonspacemuseum.com, www.collectspace.com.
4. Walter Cronkite, *A Reporter's Life*. New York: Ballantine, 1997.
5. James Kauffman, *Selling Outer Space: Kennedy, the Media, and Funding for Project Apollo, 1961–1963*. Tuscaloosa, Ala., 2009.
6. Megan Garber, "Astro Mad Men: NASA's 1960's Campaign to Win America's Heart," *The Atlantic*, July 31, 2013.

gram. Staffed largely by professional journalists, the NASA Public Affairs Office operated more like a newsroom to rapidly disseminate information to the world press. More than 3,000 reporters covered the Apollo 11 mission from the Cape and Houston, while many thousands around the world worked from home. While NASA had its share of problems with the media during the Apollo 1 fire, or with the military's Cold War habits of secrecy, its Public Affairs Office worked tirelessly from its inception through the Moon landings toward a goal of not "spinning" or "selling" the space program, but reporting it in a remarkably open way, in as close to real-time as the technology of the time allowed.

In NASA at the time of Apollo, every unit and bureau had its own public relations agenda and sense of authority, independent of headquarters, as did many individuals, including some of the dominant figures such as von Braun and, occasionally, a few of the astronauts. The main public affairs groups were severely understaffed, as no one anticipated the public's demand for information and participation. Even the tours that were offered to a demanding public at Houston's Manned Spacecraft Center and the Cape Canaveral launch site, had to be cobbled together by the staff at hand, who were at first shocked by the extent of the public's interest. Eventually, some of the tasks would be given over to a contractor (TWA, the airline, in this case). It should be said, though, that the folksiness of the staff's approach to such problems endeared it all the more to a Cold War America hungry to be in the midst of people they considered true heroes. We show how NASA's Public Affairs staff, operating with a limited budget, made the most of what they had by adopting a "brand journalism" and "content marketing" approach to educate the public through the press and broadcast media, and, especially, to educate reporters and publishers.

Where NASA fell short, the contractors, whose number included many of America's largest corporations, were only too happy to fill in with money, ideas, press releases, printed materials, press souvenirs, and even interviews when the astronauts or NASA staffers were not available. And where that was not sufficient, the media, especially in the form of television news, was happy to oblige, as it saw the space program as its own manifest destiny. Some media outlets went so far as to form consortia, to share the costs of expensive mock-ups and props, and sometimes personnel and travel, such as the pool reporters who made the long journey to cover the splashdown from a ship halfway around the world.

This greatest technological achievement of the 20th century was also a global event, as an estimated 600 million television viewers watched and listened as *Eagle* landed on the Moon and as the first human footprints marked the lunar surface. The achievement of broadcasting live television from the Moon was nearly as astonishing as landing there. Though the buildup to this moment was long, there can be no denying that the drama was epic and the dangers very great. Within the scientific community, it was understood that nearly every possible scenario required detailed preparation, ranging from anticipating physical mishaps and catastrophes to the remote likelihood that the astronauts might return to Earth carrying dangerous microbes.

On July 20, 1969, workers called in sick, children stayed home from school. Crowds gathered around televisions in department store windows and in parks where giant monitors had been set up. Newspapers and magazines, from local community papers to the ubiquitous American coffee-table staples *Life* and *Time*, along with radio and television news programs, provided unprecedented and sustained coverage to feed the

Central Park, New York City, July 20, 1969. Tens of thousands gather to watch the moonwalk on screens set up by CBS News.

TOP: *May 18, 1969. Kennedy Space Center's Deputy Director for Administration, Albert Siepert, at left on third row, points out highlights of the Apollo 10 liftoff to King Baudouin and Queen Fabiola of Belgium. Former Vice President Hubert Humphrey, wearing a cap, is seated in the second row, at right.*

ABOVE: *News reporters at Cape Canaveral launch site covering an early Mercury flight.*

voracious interest of their readers and audiences. In the drama of the event, the collective attention of millions of people became as one: focused and sellable. And thousands of organizations, from NASA's contractors to toy makers and film studios, did not miss their chance to tap into the unparalleled global wave of attention to market their products or to tell their story to an extraordinary number of potential customers. In a decade beset by controversy, protest, and strife, the adventure to the Moon was the one direction on which most people could agree, at least for a while.

The corporate marketers were also engaged on two other fronts: the business-to-business sales that might result from a company's involvement with the Apollo project, and selling goods and services to the government itself, notably the Department of Defense. Companies that were major contractors for the Apollo program, such as Boeing and Raytheon, happily talked up their lunar credentials in advertisements placed in publications read by DoD officials, such as *Aviation Week & Space Technology*. The underlying message: "If we can get America to the Moon before the Soviets, you can trust us to build your next military system." Many contractors presented their favored customers with souvenirs highlighting the company's role in the Apollo program. A father coming home with a small piece of beta cloth given to him by his sales representative from Owens-Corning Corporation, might encounter, "Wow, Dad! You work with people from the space program?" Such experiences are not soon forgotten. Also, the fact that many, if not most, of these contractors were benefitting from the escalation of the Vietnam War, the space program provided them opportunity to present themselves as fostering exploration and American pride, rather than war profiteering. Profit margins on NASA work were very lean, so much so that, during the 1960s and 1970s, many companies involved with aerospace merged and diversified their holdings and activities. (One such company, ITT, long a player in international acquisitions and mergers, acquired Sheraton Hotels and Wonderbread after losing money in aerospace.) After the Vietnam War, this strategy became critical for survival.

THE PUBLIC'S RELATIONSHIP with the space program might have played out differently. NASA's presidential mandate to be an "open program"—unlike the secretive Russian space projects—was resisted by many, especially by those from military backgrounds whose custom was to work in secrecy. The veil that covered weapons development was an aspect of the Cold War that Americans took for granted. At the beginning of America's drive into space, many citizens remained fearful in the wake of so many Soviet "firsts" in space and their perceived implications for missile-driven nuclear warfare. Yet, as the decade progressed, American society was being pushed toward greater openness and self-examination by protests against the Vietnam War, by the civil rights movement, and by counterculture voices such as the Grateful Dead. NASA policy of openness fit well with the public, especially as a counterbalance to the Dr. Strangelove military. Of course, not everything in the space program was open. The Manned Orbiting Laboratory, for example, an Air Force/NASA project, which in recent years was revealed to have been an early "spy in the sky" program, was subject to a deliberate plan of disinformation.

In the end there were enough levelheaded policy makers

who knew that many of the secrets could never be held truly secret for very long, yet to keep the program open to the public still required brave advocacy. NASA Public Affairs chief Julian Scheer, who argued passionately for open communications, became as important to the Apollo program as the first man to walk on the Moon. The first "outsider" to head up NASA PR, Scheer had been handpicked by NASA administrator James Webb to drive change. He was also a journalist and writer, and brought that valuable perspective to the job. Paul Haney, often at odds with Scheer on matters of protocol, fighting against NASA HQ's control over the public affairs mandate of the field offices, was nonetheless a passionate ally for open communications. It was Haney, in a moment of historic opportunity, who, by speaking directly with President Kennedy, won the pivotal policy victory of allowing the wide availability of mission information to journalists before a launch, rather than after, as was the policy of the military. If not for Scheer, Haney, and other NASA public affairs professionals, the American people would not have experienced the wonder of the lunar missions they paid for, at least not in the way they did. We tell their stories in these pages.

What Buzz Aldrin described as the Moon's "magnificent desolation" gradually overtook the story and the public's relationship with it. What began as epic adventure and exploration ultimately gave way to a discussion of geology, as astronaut tours became rock exhibitions. Except for the unplanned "successful failure" of Apollo 13, NASA never again captured the public's imagination after Apollo 11. What once were major news events quickly became dutiful renditions of technical details, not enough to hold the attention of broadcasters. And without that attention, space is just too expensive to fund. In our analysis, the reason humans have not been to Mars is, essentially, the result of a marketing failure.

This history begs a number of questions: How did NASA sell the space program to the American people and Congress? How did the contractors, vendors, and suppliers communicate their involvement in the program, leveraging their brand's contributions to keep the global audience focused on the Moon missions? What were its successes? And its failures?

These are the questions at the heart of *Marketing The Moon*, as we trace the story from the golden age of science fiction through the glory days of the Apollo program. And as marketing analysts do with a product whose lifecycle has passed its peak, we also trace the inevitable drop-off in interest and attention after the big goal was achieved with humans walking on the Moon. While there were, no doubt, fascinating and important robotic missions, such as the series of Mars rovers, NASA struggled to find relevance in a world with competing interests, and survives, if barely, like a company caught in an identity and brand crisis, yielding the spotlight some forty years after Apollo to marketing-savvy entrepreneurs such as Richard Branson and Virgin Galactic, Elon Musk and SpaceX, and daredevil Felix Baumgartner of the Red Bull Stratos mission, whose jump from twenty-four miles above the earth's surface, on October 14, 2012, was witnessed by the largest number of people to watch a live event on the Web. ⊙

The following is a complete list of astronauts who comprised the final flight crews in the Apollo program, as well as some others mentioned in the text. As they are often referred to by their nicknames—as the public has always known them—we offer here a formal list.

FROM THE MERCURY SEVEN

Virgil I. "Gus" Grissom
Liberty Bell 7, Gemini 3, Apollo 1

Walter M. "Wally" Schirra, Jr.
Sigma 7, Gemini 6A, Apollo 7

Alan B. "Al" Shepard, Jr.*
Freedom 7, Apollo 14

FROM ASTRONAUT GROUP 2

Neil A. Armstrong*
Gemini 8, Apollo 11

Frank F. Borman II
Gemini 7, Apollo 8

Charles "Pete" Conrad, Jr.*
Gemini 5 and 11, Apollo 12

James A. "Jim" Lovell, Jr.
Gemini 7 and 12, Apollo 8 and 13

James A. "Jim" McDivitt
Gemini 4, Apollo 9

Thomas P. "Tom" Stafford
Gemini 6A and 9A, Apollo 10

Edmund H. "Ed" White II
Gemini 4, Apollo 1

John W. Young*
Gemini 3 and 10, Apollo 10 and 16

FROM ASTRONAUT GROUP 3

Edwin E. "Buzz" Aldrin, Jr.*
Gemini 12, Apollo 11

William A. "Bill" Anders
Apollo 8

Alan L. Bean*
Apollo 12

Eugene A. "Gene" Cernan*
Gemini 9A, Apollo 10 and 17

Roger B. Chaffee
Apollo 1

Michael Collins
Gemini 10, Apollo 11

R. Walter "Walt" Cunningham
Apollo 7

Donn F. Eisele
Apollo 7

Richard F. "Dick" Gordon, Jr.
Gemini 11, Apollo 12

Russell L. "Rusty" Schweickart
Apollo 9

David R. "Dave" Scott*
Gemini 8, Apollo 9 and 15

FROM ASTRONAUT GROUP 4

Harrison H. "Jack" Schmitt*
Apollo 17

FROM ASTRONAUT GROUP 5

Ronald E. "Ron" Evans, Jr.
Apollo 17

Charles M. "Charlie" Duke, Jr.*
Apollo 16

Fred W. Haise, Jr.
Apollo 13

James B. "Jim" Irwin*
Apollo 15

T. Kenneth "Ken" Mattingly
Apollo 16

Edgar D. "Ed" Mitchell*
Apollo 14

Stuart A. "Stu" Roosa
Apollo 14

John L. "Jack" Swigert, Jr.
Apollo 13

Alfred M. "Al" Worden
Apollo 15

OTHER ASTRONAUTS MENTIONED

Donald K. "Deke" Slayton
Director, Flight Crew Operations
Apollo-Soyuz Test Project

Sally Ride
STS 7 and 41-G

* *Apollo astronauts who walked on the Moon.*

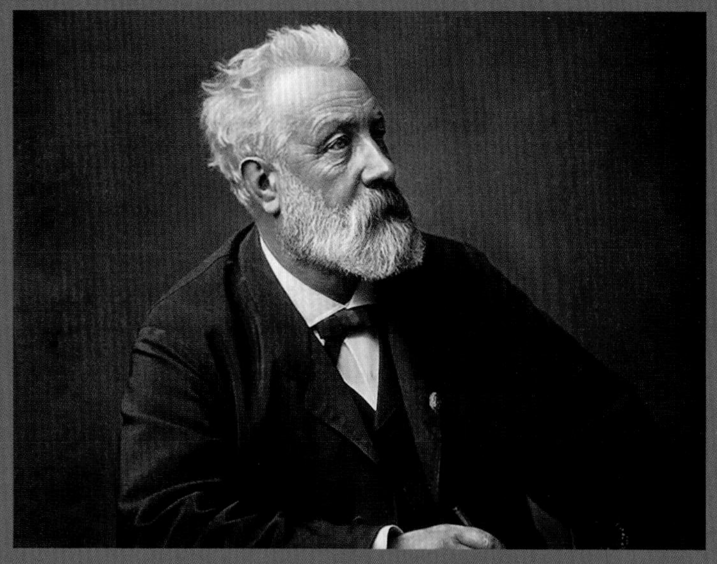

LUNAR VISIONS AND VISIONARIES. *Clockwise from upper left: Poster for Fritz Lang's 1929 German silent film* Frau im Mond *("The Woman in the Moon"), on which pioneering rocket science advocates Hermann Oberth and Willy Ley acted as consultants; Jules Verne; a frame from cinema's first Moon landing as depicted in Georges Méliès's 1902 film* Voyage dans la Lune, *loosely based on Verne's novel and recreated in Martin Scorsese's 2011 film* Hugo; *the character played by actress Gerda Maurus in Lang's* Frau im Mond *conducts some lunar cinematography; the binding of an early English language translation of Verne's 1865 novel.*

A Modern-Day *Columbia*: Fiction Makes a Giant Leap

"THIS WILL BE OUR LAST TV show," wrote Michael Collins about the final live television transmission to the world from the crew of Apollo 11, broadcast just a few hours before splashdown. It was Wednesday night on the East Coast of the United States, July 23, 1969. After the intense drama and elation over the successful lunar landing and liftoff, the return trip had been almost routine for the crew. Command module pilot Michael Collins, along with moonwalkers Neil Armstrong and Buzz Aldrin, decided to use their final broadcast to send a global message to the world. Unlike the earlier transmissions, on this occasion the crew had at least an hour to contemplate and prepare what they wanted to say.

"Although we didn't want the bloody thing on board in the first place, this time we are going to try and make the goddamn tube work for us," wrote Collins, reflecting on that historic moment. "We [were] using this last opportunity to make our statement."[1] The Apollo 11 mission dominated print, radio, and television news on Sunday, July 20th, and Monday, the 21st, as the major networks presented continuous coverage and newspapers printed special editions and features. By Wednesday evening, however, the networks and newspapers had returned to their regular programming and editorial schedules, with only occasional updates and reflection pieces.[2]

At CBS, veteran newsman Walter Cronkite and former astronaut Wally Schirra were preparing to go on the air shortly after 7:00 P.M. EST to cover the final Apollo 11 broadcast. The network, which had called upon the services of nearly one thousand individuals during the weekend marathon telecast of the lunar landing, was now relying on less than a quarter that number to cover the remainder of the flight.[3]

As millions watched their televisions, they saw the color transmission begin with Neil Armstrong. The camera was focused on the round Apollo 11 mission insignia patch on the breast pocket of Armstrong's flight suit, and the lens remained trained on it as he spoke his introduction.[4] The image was powerful and deliberate: it depicted an American eagle, spread-winged over the lunar surface, with an olive branch clutched in its talons. It was a clear, symbolic restatement of the message inscribed on the lunar module's plaque: "We came in peace for all mankind." In his last message to the world from space, Neil Armstrong framed their recent achievement and its pledge of peace by alluding to its literary predecessor.

> "Good evening. This is the commander of Apollo 11. A hundred years ago, Jules Verne wrote a book about a voyage to the Moon. His spaceship, *Columbia*, took off from Florida and landed in the Pacific Ocean, after completing a trip to the Moon. It seems appropriate to us to share with you some of the reflections of the crew as modern-day *Columbia* completes its rendezvous with the planet Earth and the same Pacific Ocean tomorrow. . . ."[5]

From a public-relations perspective, this was a masterful message coming at the moment when public interest in manned space exploration was at its apex. By linking the flight of Apollo 11 with its famous literary predecessor, Armstrong paid tribute to the power of the human imagination—reminding the world that a seemingly impossible dream can spark curiosity and motivate others to make it a reality.

This was not the first time that Armstrong had referenced Jules Verne. At the July 5, 1969, crew press conference, he unveiled the names of their spacecraft—and explained the rationale behind selecting *Eagle* as the name for the lunar module and *Columbia* for the command module. "*Columbia* was the name Jules Verne picked for a fictional spacecraft which journeyed to the Moon a century ago," Armstrong told them.[6] The press loved the allusion. "Jules Verne Foretold Apollo 11 Flight Some 104 Years Ago" was the title of the article written by the Associated Press's Harry Rosenthal; United Press's H.D. Quigg wrote "Jules Verne's Fictional Lunar Trip Nears Truth." Both appeared in hundreds of newspapers

1. Michael Collins, *Carrying the Fire: An Astronaut's Journey*. New York: Farrar, Straus & Giroux, 1974, p. 430.
2. Other news was capturing the headlines: on Chappaquiddick Island, in Massachusetts, reporters were investigating conflicting accounts of a fatal automobile accident that had occurred during the weekend of the Moon landing, in which Senator Edward Kennedy, the presumed 1972 Democratic presidential front-runner, had been involved.
3. William David Compton, *Where No Man Has Gone Before: A History of Apollo Lunar Exploration Missions*. Collingdale, PA: Diane Publishing Co., 1989, p. 142
4. Apollo 11 original video footage. NASA, 1969.
5. *Apollo 11: Technical Air-to-Ground Voice Transcription*. NASA: July 1969, p. 588.
6. "Astronauts Meet Press: Moon Trip Draws Near," *The Hutchinson News*, Sunday, July 6, 1969.

THE THREE FATHERS OF MODERN ROCKET SCIENCE. Left to right: The American Robert Goddard photographed at Clark University in 1924. The cover of Hermann Oberth's 1923 book Die Rakete zu den Planetenräumen *("By Rocket into Planetary Space"), which the German author self-published after it was rejected for a doctoral dissertation. A Russian ruble coin commemorating Konstantin Tsiolkovsky.*

around the country. Reporters seemed to enjoy cataloging the similarities between the actual flight of *Columbia* and the fictional Moon mission of Verne's *Columbiad* of 1865.

"Astronautics is unique among all the sciences because it owes it origins to an art form," wrote astronomical artist Ron Miller in an essay, "Spaceflight and Popular Culture." "Long before engineers and scientists took the possibility of spaceflight seriously, virtually all of its aspects were explored first in art and literature, and long before the scientists themselves were taken seriously, the arts kept the torch of interest burning.... No one had considered the actual technological problems of space flight until Jules Verne."[7] Prior to *From the Earth to the Moon*, Miller points out that all tales of space travel had been fantasies of one type or another. After Verne, "the possibility of spaceflight was instantly transformed from the realm of the fantastic ... for the first time, the problem of space travel had been put on a firm, mathematical and technological basis."[8]

The recognized "founding fathers" of modern rocketry, Robert Goddard in the United States, Hermann Oberth in Germany, and Konstantin Tsiolkovsky in Russia, drew inspiration from fictional tales of interplanetary travel, and even wrote science fiction themselves.[9] "My interest in space travel," Tsiolkovsky wrote, "was first aroused by the famous writer of fantasies Jules Verne. Curiosity was followed by serious thought." Oberth read Verne's lunar voyages so many times he finally knew them by heart, and Goddard read and re-read *From Earth to the Moon* and wrote comments and corrections in the margins.[10]

It was only after articles about Nazi Germany's V-2 program appeared in the press that rocket technology entered the public's consciousness as a reality. A few years earlier, the *New York Times* had gone so far as to publish an editorial stating there could be no doubt that a rocket operating in the vacuum of outer space was "absurd."[11] But in the wake of the V-2 and the explosion of the atomic bomb, previously existing scientific and technological certainties were open to revision. Perhaps human space travel—even extraterrestrial life—was indeed plausible. It's hardly coincidental that the first widely reported sightings of flying saucers occurred in 1947. This growing awareness of a new and uncertain future is indicated in a 1949 George Gallup poll surveying attitudes about the half-century ahead. Looking ahead to the year 2000, 83% of Americans foresaw a cure for cancer, and 63% thought there would be atomic trains and airplanes.[12] The possibilities seemed unlimited, and the imaginations of everyday Americans were deeply engaged. So were their fears. Some of the first articles published about space during the early days of the Cold War defined it primarily as the location of humanity's next battleground.

Far more often, though, the exploration of outer space was

7. Ron Miller, "Spaceflight and Popular Culture" in *Societal Impact of Spaceflight*. NASA, 2007, p. 501.
8. Ibid.
9. Neil McAleer, *Odyssey: The Authorized Biography of Arthur C. Clarke*. London: Victor Gollancz, 1992, p. 179.
10. Verne's influence on Tsiolkovsky, Oberth and Goddard are noted by Howard E. McCurdy, *Space and the American Imagination*, Smithsonian Institution, 1997, pp. 15–16.
11. Unsigned editorial, January 13, 1920.
12. Cited in Michael J. Neufeld, *Von Braun: Dreamer of Space, Engineer of War*. New York: Knopf, 2007, pp. 266–267.

popularly described using language and archetypes borrowed from America's own frontier mythology. Here was a new wilderness to be explored and conquered. The high frontier appeared the inevitable next chapter in the nation's epic, a destiny that would challenge American innovation and technology, and limn the American character of coming generations. Throughout the first half of the 20th century, American popular culture was dominated by the Western, with stories widely disseminated in print, on radio, in cinemas and, eventually, on television. Then, at the century's midpoint, there began a very gradual shift in the popularity of storytelling genres, and a new reinterpretation of the American myth. The iconic frontier of the historic past that was defined in the Western genre entered its maturity and declined in popularity until it nearly disappeared from theater and television screens during the early 1970s. Its decline coincided with a gradual and slowly growing popular enthusiasm for science fiction entertainment, which, during its formative years, relied heavily on familiar mythic Western archetypes. Sometimes this collision was overt. For example, when Desilu studios attempted to sell the NBC network the original concept of *Star Trek*, in the early 1960s, it conveniently described the series as "*Wagon Train* to the stars," a direct comparison to a popular ABC network television Western of the time.[13]

Hollywood's infatuation with space-related entertainment began in 1950 with the release of *Destination Moon* and *Rocketship X-M,* among the first science fiction films from a decade that came to be partly defined by them. Their box office success encouraged producers to chase after audiences that appeared eager to see escapist stories set in unearthly realms or depicting the possibility of extraterrestrial life. Though more often than not, these tales tapped into the collective fears of the Cold War. In the films that immediately followed, space aliens were either overtly belligerent, as in *The Thing from Another World* (1951), *The War of the Worlds* (1953), or perceived as threats, as in *It Came from Outer Space* (1953) and *The Day the Earth Stood Still* (1951). One of the few films of the era that tried to evoke the excitement of space travel in the near future, *Conquest of Space* (1955), was saddled with a poor screenplay and an unenthusiastic studio. Its failure at the box office scared away American producers from further realistic depictions of space travel for thirteen years, until *2001: A Space Odyssey,* was released a few months before man visited the Moon.

Nevertheless, the cinema—and to a lesser degree television—had firmly established outer space as a realm of romance and adventure. As larger and larger audiences let their imaginations soar among the stars, merchandisers seized opportunities to create products capturing the excitement and allure of extraterrestrial travel. The first wave of space-related toys came during the Depression and coincided with the popularity of the *Buck Rogers* and *Flash Gordon* serials. However, as television sets proliferated during the 1950s and Hollywood released a sudden onslaught of science fiction films, space-themed toys became more and more common, though never eclipsing the ubiquity of the frontier, cowboy, and Native American merchandise popular during the Eisenhower years. An early indication of shifting tastes came in 1951, when Post Cereals dropped their advertising on network television's first Western series, *Hopalong Cassidy,* to sign a $10 million, five-year contract to sponsor *Captain Video*. At that time an article reported that there were more than 284 tie-in items marketed to fans of cowboy Roy Rogers, while licensing agreements for merchandise related to the *Tom Corbett, Space Cadet* television series numbered no more than 100. (In all, the total market for television show tie-in merchandise in 1951 was estimated to be more than $250 million.)[14]

The increasing proliferation of space toys, clubs, games, activity books, trading cards, uniforms, and other products did much to generate curiosity about the cosmos and establish outer space as a vivid background for imaginary play and youthful wonder. Indeed, these space adventures—the *Buck Rogers* and *Flash Gordon* serials of the 1930s and 1940s, and early television's *Tom Corbett, Captain Video, Space Patrol, Dimension X,* and others—and the merchandising tie-ins associated with them comprised a decisive and influential moment in the saga of marketing space exploration to the population at large.

Space Patrol, which debuted in 1950 and continued into 1955 (its radio incarnation was broadcast from 1952 to 1955), was at first directed toward children, but over time developed a sizable adult audience. It followed the 30th-century adventures of Commander-in-Chief Buzz Corry of the United Planets Space Patrol and his young sidekick Cadet Happy, as they

13. Talk given by former Desilu producer Oscar Katz at the first *Star Trek* convention, January 1972.
14. Ernest La France, "Watch Out, Here Come the Space Men!" *Parade Magazine*, October 14, 1951, pp. 20–21.

Space Patrol

Space Patrol, the first space-themed television series to become a national sensation, is enshrined in the annals of marketing for one of the most memorable promotions in broadcasting history. A children's adventure series very much in the spirit of the *Flash Gordon* and *Buck Rogers* serials of the 1930s, *Space Patrol* was broadcast nationally on the ABC network in 1951 after first appearing exclusively on the West Coast a year earlier. Its stories followed the exploits of Commander-in-Chief Buzz Corry of the United Planets as he and his compatriots traveled through time and battled interplanetary villains. As science fiction, *Space Patrol* was a throwback to the space operas of the early pulp magazines, yet when televised nationally, its audience mushroomed in size and included many adults. Shortly after, it spawned its own popular radio series, as well.

The program's primary sponsor was the Ralston-Purina Company, which heavily promoted its breakfast cereals, Rice Chex and Wheat Chex, in *Space Patrol* advertising on television and radio, on records, in comic books and comic strips, and in point-of-purchase store displays. Nearly every episode included an offer for a toy premium or a club membership that could be obtained only by mailing in the required proof-of-purchase cereal box tops. Sometimes the featured premium was also integrated within the *Space Patrol* storyline to create viewer excitement and generate sales. (For example, in the radio episode "The Prisoners of Pluto," listeners learned that, for a Chex box top and 35¢, they could obtain a *Space Patrol* pocket Project-o-Scope, "complete with bulb, batteries, and film," which also served as a critical prop in that episode .)

But no other marketing gimmick compared to *Space Patrol*'s "Name the Planet Contest," often cited as the biggest television promotion of its period. Developed by the Gardner Advertising Company and launched in 1953, the contest featured a first prize that immediately captured the attention of all *Space Patrol* fans: an actual five-ton, thirty-five-foot rocket club house, a reproduction of the *Terra IV* battle cruiser from the show. To enter, contestants were required to submit an album of plastic space coins with their entry form (each coin required the purchase of a box of cereal). The full-size *Terra IV* rocket clubhouse and the flatbed truck on which it rested were originally created by the Gardner Agency as a promotional attraction to advertise Ralston Purina products. The "rocket" appeared at fairgrounds, parks, and supermarket openings, pioneering what was then a new and innovative marketing strategy. At such events, children and their parents could tour the space ship, with a box top from a package of Ralston cereal as their ticket.

protected Earth and its allies against a variety of villains, a number of whom had curiously German and Russian accents. Their base was a space station. The idea of such a space-based peacekeeping force developed some currency in certain U.S. military circles. The earliest comprehensive plan was articulated in a 1963 address by Lt. Gen. James Ferguson, the Air Force deputy chief of staff for research and development, which made direct reference to the television series. Though Ferguson's vision of a military Space Patrol never got under way as planned, it took on another life in the form of the Manned Orbiting Laboratory project, which was announced in December 1963. The MOL was to be an orbital laboratory for forty-day missions, utilizing a Gemini-style capsule for Earth return. After five years of planning, development, and astronaut training, the project was canceled in 1969.

It wasn't all space cadets, alien invasions, and flying saucers during the 1950s. The fantastic literature that had found a loyal readership during the age of pulp magazines had matured. By the 1940s, the generation that had grown up reading tales of spacemen blasting bug-eyed monsters with ray guns, started writing their own, more realistic form of "hard science fiction," in which plausible technology and incontestable scientific laws were given precedence over authorial license. Many among the new band of writers, including Isaac Asimov, John W. Campbell, Hal Clement, and Arthur C. Clarke, were working scientists or had extensive training in astrophysics, biochemistry, or other specialized disciplines. Additionally, a few science fiction authors began appearing in the better-paying, higher-profile mass circulation magazines, known as the "slicks." In particular, Ray Bradbury, a science fiction magazine regular, began publishing stories in *The Saturday Evening Post* and *Collier's* in 1950, making him one of the most widely read contemporary authors of fantastic literature in the country. Leading American book publishers, having previously relegated science fiction to smaller, specialty houses, were now aware of the growing market and began releasing titles in hardcover and inexpensive paperback editions.

The early years of science fiction wove a rich and vivid cultural tapestry against which the future Apollo space program, as well as its vendors and suppliers, had to contend. For anyone selling products associated with the American manned space flight, this cultural legacy offered wonderful opportunities to access the collective fantasies and excitement associated with space travel. For many in NASA's Public Affairs Office, it appeared to contribute to a growing sense of apathy about the manned lunar program as the general public became aware that the realities of spaceflight differed greatly from the romanticized versions envisioned in magazines, books, and on cinema screens.

The Merging of Science Fiction and Fact: The *Collier's* Series

The increasing influence of science fiction literature, and Hollywood's expanding interest in stories set in space, prompted the public to consider extraterrestrial travel in imaginative ways they had not done before. Yet most space historians point to an additional cultural moment from the spring of 1952 as having spawned a greater and longer-lasting impact. This was the publication of the March 22nd issue of *Collier's* magazine with a cover illustration depicting a rocket-plane in orbit high above Earth accompanied by the emphatic headline, "Man Will Conquer Space Soon." This was the first of a series of eight issues *Collier's* published over the next two years that persuasively made the case for manned space exploration to the Moon and Mars within the foreseeable future. Each issue featured accessibly written articles by leading authorities in their fields, including rocket pioneer Wernher von Braun, science writer and space travel promoter Willy Ley, astronomer Fred L. Whipple of the Harvard Observatory, and others. As important as the text were the accompanying dramatic illustrations by artists Chesley Bonestell, Fred Freeman, and Rolf Klep, which looked like nothing seen previously in a popular, general-interest magazine.

The first appearances of these special issues of *Collier's* are still vividly remembered by impressionable minds of the era. Albert A. Jackson, an astrophysicist and planetary scientist, who worked as an Apollo crew training instructor at Manned Spacecraft Center, was eleven years old when he first set eyes on the cover of the October 18, 1952, issue of *Collier's* featuring the headline "Man on the Moon: Scientists Tell How We Can Land There in Our Lifetime." He was captivated, yet shocked, at the design of the spaceship on the cover. It didn't look right. "Where was the bullet shape? The fins? The needle nose? This was not right! This was supposed to be a space ship! It was ugly! Yet, that lighting, the color, that splash of molten rock! The detail! How could something so ugly catch my imagination? How could it be so real? I took that issue to my room. . . . That week I must have read that issue twenty times!"[15] And it changed the course of his life.

These eight issues of *Collier's* also affected the future course of American space exploration. A once widely read magazine with an estimated circulation of nearly three million, *Collier's* was experiencing financial challenges and declining readership during the early 1950s. Advertising was already beginning to migrate to television, and *Collier's* was hurt by competition from other successful, illustrated, mass-circulation magazines, particularly *Life, The Saturday Evening Post*, and *Look*. Fortuitously, two *Collier's* reporters were among an estimated 200 attendees to the First Annual Symposium on Space Travel held on Columbus Day, 1951, at New York's Hayden Planetarium. (Both Willy Ley, who was also coordinator of the conference, and Fred Whipple were among the speakers.) The managing editor of *Collier's*, Gordon Manning, assigned an associate editor, Irish-born journalist Cornelius Ryan (later famous as the author of the bestselling chronicle of World War II's Normandy Invasion, *The Longest Day*), to determine whether a feature article might be developed from this material. Initially, Ryan was a space-travel skeptic, but after listening to von Braun, Whipple, and their associates he became a believer. In turn, Ryan persuaded Manning that *Collier's* should convene its own symposium on humanity's future in space: a conclave of specialists that would not only generate a series of unique feature articles, but would also have potential to spark news and publicity, lure back advertisers, and revive flagging circulation.[16]

Collier's heavily promoted the March 22, 1952, issue with advertising, metropolitan department store window displays, and national media. The publicity push was launched with the distribution of 2800 time-sensitive press kits, clearly marked on the cover "For *exclusive* use of press, radio and television. For release, 7 P.M. EST, Thursday, March 13th. Please guard against premature release." The kit contained photos and a press release that emphasized the Cold War defense aspects of a von Braun-designed American space station described and pictured in the issue that went on sale Friday, the 14th. According to the release, "In a two-page editorial, the editors of *Collier's* state that in the hands of the West, such a space station, permanently established beyond the atmosphere, 'would be the greatest hope for peace the world has ever known.' The editorial emphasizes that the scientific data presented by Dr. von Braun and five other leading scientists in *Collier's* serve as an urgent warning that the United States must immediately embark on long-range development to secure for the West space security. If we do not, someone else will. That somebody else very probably would be the Soviet Union." Possible trips to the Moon from the orbiting platform merited only a small, secondary note within *Collier's* release.[17]

Newspaper advertisements appearing the morning of March 14 featured a picture of the circular space station orbiting Earth above a headline "We can conquer space in 10 years—*and guarantee peace forever!*"[18] Von Braun was booked on national television and radio to promote *Collier's,* and had already appeared the previous evening on NBC's *Camel News Caravan,* shortly after the 7 P.M. news embargo had been lifted.[19] On the air, von Braun used visual aids to illustrate his points: detailed scale models of the launch vehicle and space station constructed by his rocket team. Within the next twenty-four hours he was interviewed on NBC's *Today Show, Garry Moore* on CBS, and even made an appearance on ABC's *Tom Corbett, Space Cadet*.[20] Although von Braun had previously spoken to select groups around the United States about the future of space travel, this was his introduction to the American mass public. His face and accented speech would soon become familiar presence on television during the coming quarter century. Von Braun later referred to the *Collier's* publicity effort as "by far the greatest public advertising campaign for spaceflight . . . the world has ever seen."[21]

Much of the content for the *Collier's* series was developed during an earlier attempt by von Braun's to promote manned space travel by writing a science fiction novel about a trip to Mars. In conjunction with his novel, written during 1947 and 1948, he and his rocket team worked out the essential technical details for an interplanetary voyage, a wealth of information later popularized in the magazine articles. One of von Braun's associates confessed that, in 1947, few "believed there would be a social or economic setting in the near future that would make the concept real," but they had fun playing with the idea.[22]

Promoting national defense in a future battleground in

15. Albert A. Jackson, "The Ugly Spaceship and the Astounding Dream," *AIAA Houston Section Horizons,* April 2002, pp. 3, 14.
16. Background about the Collier's series is found in Randy Liebermann's invaluable essay "The Collier's and Disney Series" published in *Blueprint for Space: Science Fiction to Science Fact*, Washington, D.C., Smithsonian Institution Press, 1991.
17. "News from Collier's" press release dated March 13, 1952.
18. Newspaper advertisement that appeared within the Collier's press kit.
19. Michael J. Neufeld, op. cit., p. 258.
20. Ibid.
21. Harlen Makemson, op. cit., p. 13.
22. Albert A. Jackson, "The Conquest of Space," *AIAA Houston Section Horizons,* March/April 2012, p. 47.

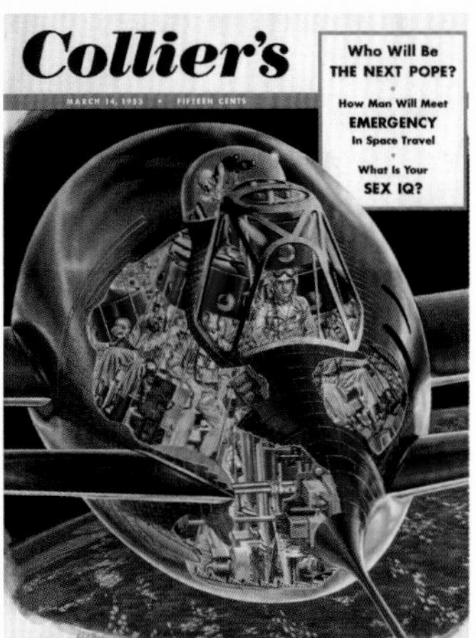

The influential space exploration issues of Collier's published between 1952 and 1954. All of the cover illustrations are by Chesley Bonestell, with the exception of the cutaway image showing the interior of a manned spacecraft painted by Fred Freeman (top right), and the photograph showing the "World's First Space Suit" (center bottom).

The Collier's Visionaries

No artists did more to capture the romance and fascination of manned space exploration during the formative years of the space program than the three illustrators chosen by *Collier's* art director William Chessman to graphically accompany the revolutionary articles that appeared between 1952 and 1954.

Chesley Bonestell (1888–1986) was by far the best known of the trio, and his astronomical paintings showing scenes of Saturn viewed from its moons created a sensation when *Life* magazine published them in 1944. During the 1940s and early 1950s, Bonestell worked in Hollywood, creating fantastic photorealistic images such as the exterior of Xanadu in Orson Welles's *Citizen Kane* (1941) and Howard Roarke's skyscrapers in King Vidor's *The Fountainhead* (1949). With Willy Ley, Bonestell collaborated on the speculative science book, *The Conquest of Space* (1949). The year before, his image of a rocket entering Earth orbit announced "Man Will Conquer Space Soon." Bonestell was commissioned by *Collier's* to create harrowing, aerial-view depictions of Manhattan and Washington, D.C., under nuclear attack. It is hard to overestimate the influence of his paintings on all future depictions of man's destiny in space. Both a crater on Mars and the asteroid 3129 Bonestell are named in his honor.

Fred Freeman (1906–1988) ran away from his home in West Newton, Massachusetts, at age 16 to become an illustrator in New York City. During World War II, he was a Naval Reserve lieutenant commander, skippering three different ships and seeing action in the Pacific. His wartime experience led to a commission from the Naval Institute Press, which needed detailed, dramatic illustrations of men in confined spaces for two books on submarines and destroyer operations during World War II. In 1961, he illustrated Arthur C. Clarke's novel *A Fall of Moondust*. Of the *Collier's* artists, von Braun was closest to Freeman, who illustrated his book *First Men to the Moon* and provided pen and ink drawings for von Braun's fictional story "Life on Mars," serialized in *This Week*.

Rolf Klep (1904–1981). By 1952 Klep was known as one of the nation's finest technical illustrators, familiar to readers of *Fortune* magazine for his detailed renderings of everything from a cutaway view of a factory layout to an exploded diagram showing the components of the latest Army Air Force bomber. During his early career, Klep often worked in a combination of pen-and-ink and watercolor, but later became a master of the airbrush, which gave much of his technical work smooth mechanical precision. During World War II he was in charge of graphic art production for the Office of the Chief of Naval Operations, and he saw action in the Pacific and Atlantic theaters. After his retirement in 1962, he founded The Columbia River Maritime Museum in Astoria, Oregon.

Wernher von Braun's *Project Mars*

Robert Goddard, Hermann Oberth, and Konstantin Tsiolkovsky weren't the only rocketry pioneers whose destiny was influenced by reading early science fiction adventures. As an adolescent during the 1920s, a young Wernher von Braun also devoured the novels of Verne, H. G. Wells, and Kurd Lasswitz, whose 1897 adventure *Auf zwei Planeten* ("On Two Planets") sparked von Braun's lifelong interest in the exploration of Mars. Far less well known is the fact that, like Goddard, Oberth, and Tsiolkovsky, von Braun wrote science fiction, authoring a novel, *Das Marsprojekt*, while he and his team of German rocket scientists were languishing without official work under the supervision of U.S. Army at the White Sands Proving Grounds in New Mexico. Von Braun believed that a popular science fiction novel about an expedition to Mars might inspire others the same way the works of Verne, Wells, and Lasswitz had motivated his own career. As von Braun wrote his novel in late 1947 and early 1948, he called upon the expertise of many of his associates to conceptualize and detail technical aspects of the imaginary interplanetary mission. He later summarized the technical details in a scientific appendix that would appear as a supplement to the novel's manuscript. Sadly for von Braun, no American publisher expressed interest in the novel, though a reworked and expanded version of the appendix alone appeared in Germany as a book in 1952. (An American edition appeared a year later, published as *The Mars Project* by University of Illinois Press.) This was to be von Braun's lone attempt at writing science fiction. Ironically, the novel's detailed technical research led directly to the influential series of *Collier's* magazine articles and the subsequent Disney television films they inspired. Revised portions of von Braun's foray into fiction appeared under the title "Life on Mars," published in three serial installments in *This Week,* a magazine that came with many American Sunday newspapers. Each installment featured illustrations by artist Fred Freeman, who also created many of the memorable and dramatic images that accompanied the *Collier's* series. Twenty-six years after his death, von Braun's entire novel was finally published in an English-language edition under the title *Project MARS: A Technical Tale*.

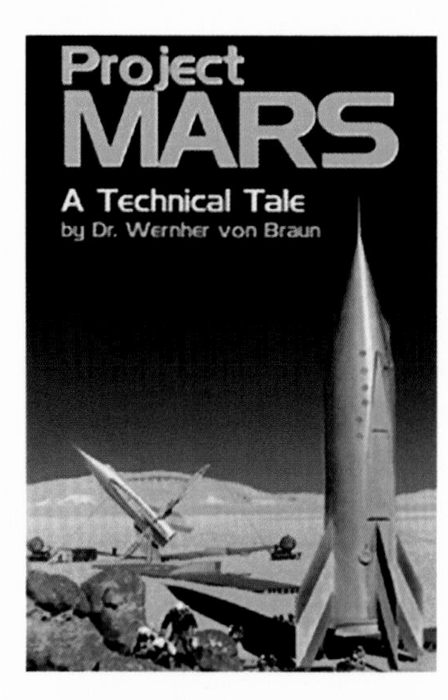

Two worlds meet:
The Martians were polite, the Earthmen wary

Pillow-speakers taught Martian

Cables secured the return rocket

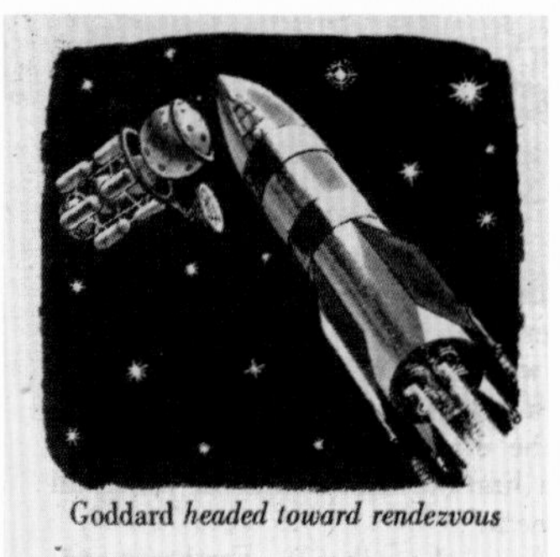
Goddard *headed toward rendezvous*

McKay braked the tractor as Sterling exclaimed: "I swear there's a temple up there!"

space served as *Collier's* publicity hook. But national security was a minor portion of the content in *Collier's* "Man Will Conquer Space Soon" issue, which was primarily focused on the technicalities of reaching orbital velocity, building a permanent station in space, extraterrestrial law ("Who Owns Space?"), and how humans would adapt to weightlessness and the vacuum of space.

As unique and visionary as these articles were when they appeared within the pages of a mass-circulation magazine, they would have had far less impact had they been isolated from the invaluable visualizations by Bonestell, Freeman, and Klep. The *Collier's* space paintings are now considered among the most influential images of the early space age, and served as the basis for countless subsequent book and magazine illustrations. More than a half-century later, this artwork still evokes awe and wonder, and they are indisputably among the seminal artifacts that sold dreams of space to a generation of impressionable Americans. Bonestell, living and working in California, and Freeman and Klep, based in New York, all collaborated with the *Collier's* authors, often working from original engineering diagrams supplied by von Braun.[23] The March 22, 1952, issue of *Collier's* made the American public consider the realities of space travel in a way the media had never done previously. Films like *Destination Moon* and *Rocketship X-M* appealed to audiences as escapist entertainment, but the twenty-five pages devoted to space travel in *Collier's* were sober and persuasive. When George Gallup once again surveyed public expectations regarding scientific and technical advances during the next fifty years, the increased percentage of Americans who believed it possible that "men in rockets will be able to reach the Moon in the next fifty years" was notable. The 1949 poll indicated that 15% thought it plausible. Six years later this segment had grown to 38%, and the *Collier's* series is believed to have been a primary factor in this increase.

During the next two years, *Collier's* published seven more space-themed issues building on the first. In the October 18 and October 25, 1952, issues—the former being the one that captured the imagination of eleven-year-old Albert A. Jackson—von Braun and Ley made their case for the first lunar voyage. Once again, Bonestell, Freeman, and Klep provided invaluable illustrations, and von Braun engaged in additional television and radio interviews—a second publicity blitz.

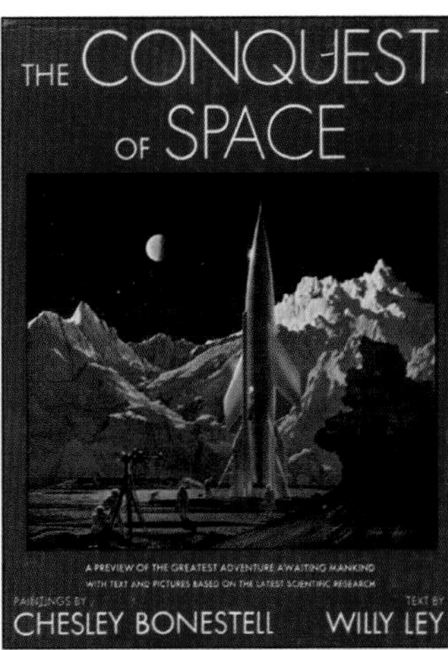

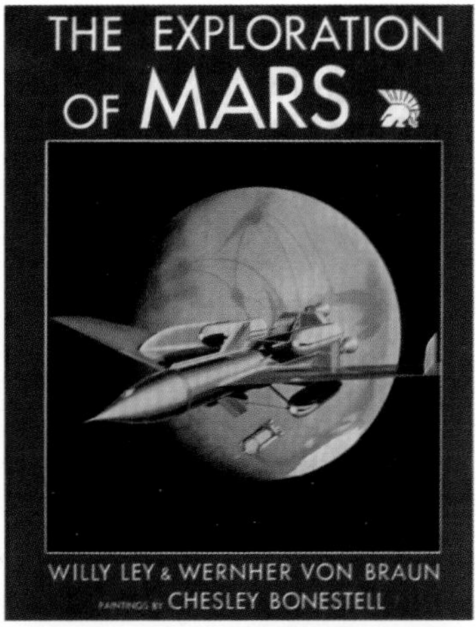

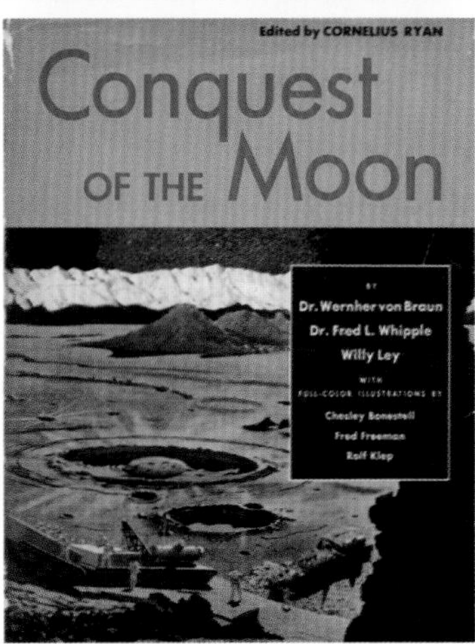

Viking Press, publisher of Willy Ley and Chesley Bonestell's 1949 book, *The Conquest of Space,* swiftly contracted for two similarly designed large-format volumes expanded from the *Collier's* series: *Across the Space Frontier*, published in September 1952 (timed for the Christmas book season), and *Conquest of the Moon,* in 1953. Any child infected by the romance of space travel during the 1950s remembers these books vividly—they were in constant demand at public libraries throughout the United States during the next decade. A fourth Viking Press book, *The Exploration of Mars,* a long-planned sequel to *The Conquest of Space* (contracted years before the *Collier's* series was published), was released in 1956, timed for publication when Mars reached its perigee with the Earth.

During late winter 1953, *Collier's* devoted twenty-seven pages in three subsequent issues exploring the subject of the first astronaut corps: how to select the initial group of men to

23. Randy Liebermann, op. cit, pp. 136, 139.

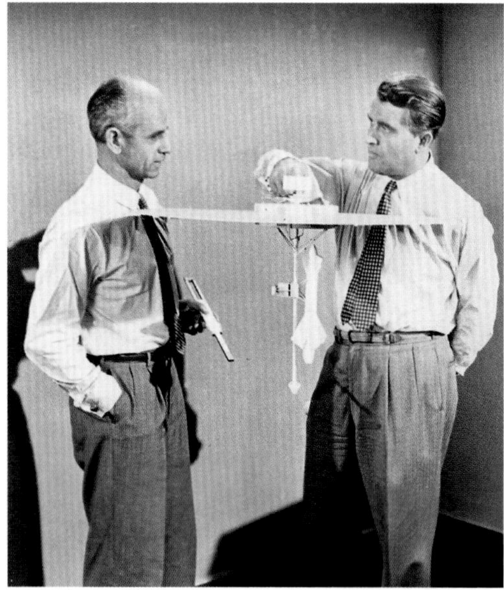

Walt Disney, Wernher von Braun, and "Man In Space," 1955. *From left to right: Walt Disney; Ernst Stuhlinger and Wernher von Braun appearing in Disney's "Mars and Beyond"; von Braun publicity shot; from Disney's introduction; Disney and von Braun; Heinz Haber, von Braun, and Willy Ley. Below: Dell comics version.*

fly in space, as well as the physical and psychological challenges they would encounter. Four months later, a short article by von Braun and Ryan detailed plans for the first space station, an experimental platform in which remote monitors would follow the progress and health of a crew of trained Rhesus monkeys during a mission of sixty days.

The *Collier's* series concluded nearly a year later with the April 30, 1954, issue devoted to von Braun's personal obsession: a trip to Mars. Astronomer Fred Whipple wrote about the possibility of life on the Red Planet, while von Braun and Ryan outlined plans for an eight-month voyage by a fleet of ten spacecraft. Originally planned for publication nearly a year earlier, this final installment was delayed when the editorial focus of the series shifted in response to reader interest about the first corps of astronauts. This final installment was released so long after the preceding seven, *Collier's* management concluded little was to be gained by mounting an additional publicity push. Their reluctance to promote this long-awaited issue was likely indicative of the health of the magazine as well. *Collier's*, the venerable weekly that first appeared in 1888 and became famous at the turn of the century for pioneering investigative journalism by writers such as Upton Sinclair and Ida Tarbell, ceased publication less than three years later. Yet, without those influential eight issues published during the magazine's final years, the history of the space age would likely have been very different. And, if not for *Collier's*, the occurrence of the second major media event of the 1950s that sold space to the American public would never have happened.

Disney's *Man in Space* and Tomorrowland

The second turning point in the 1950s' marketing of the Moon came when American families switched on a relatively new living room appliance on the evening of March 9, 1955. An audience estimated at nearly forty-two million—nearly a third of the population of the United States—turned the dial to the ABC television network to view an entertaining prime-time program about the history of rocketry that climaxed with an animated trip to a manned orbiting space station of the near future.[24]

The two personalities directly responsible for bringing that broadcast to the airways were skilled marketers, supremely adept at the art of persuasion and communicating to the masses. Not surprisingly, Wernher von Braun was involved once again. His partner this time was Walt Disney.

"Man in Space" was shown as part of *Disneyland,* Walt Disney's first foray into prime-time television, and an extension of his brand reaching directly into American homes. In addition to recycling Disney's growing library of films and animation, the weekly program served as a novel way to promote his first theme park, which opened in Anaheim, California, on July 18,

1955. Each episode of *Disneyland* centered around one of the four themed locations within the Disneyland resort: "Frontierland," "Fantasyland," "Adventureland," and "Tomorrowland." Realizing that his studio had little existing content to showcase on *Disneyland* under the "Tomorrowland" banner, Disney assigned one of his key studio employees, veteran animator Ward Kimball, to develop some program ideas imagining the world of the future.[25]

Disney imparted to Kimball one important condition: the programs must be deeply rooted in science. To distinctly set apart the "Tomorrowland" segments from other television space dramas, Disney coined the term "science-factual." Kimball had been fascinated by the recent *Collier's* series and called upon the expertise of three of its leading contributors: space advocate and writer Willy Ley, aerospace medicine consultant Heinz Haber, and von Braun, now leader of the development team at the Army's Redstone Arsenal. With a budget of more than $1 million, Kimball produced three films in the *Disneyland* series, with "Man in Space" followed by "Man and the Moon" (1955) and "Mars and Beyond" (1957), all inspired by the articles that had appeared in *Collier's* a few years earlier.

Now an experienced media veteran following the *Collier's* publicity campaign, von Braun was comfortable on camera and knew how to sell his message. In "Man in Space," he appeared on screen presenting his outline for a four-stage passenger rocket that, he suggested, if backed by an organized and well-supported space program, "could be built and tested within ten years."[26] Working with the Disney artists, von Braun ensured that the depiction of the realistic animated launch sequence and the space vehicles shown in orbit were scientifically accurate. According to Ernst Stuhlinger, a von Braun associate who was also a technical consultant on "Man in Space," the pair would often fly out to California on business trips to visit defense contractors, and then meet with Disney artists and producers in energetic all-night sessions.[27]

With his appearances on *Disneyland,* von Braun enhanced his growing national profile, despite some personal unease with the public attention. In contrast to Willy Ley, who projected the persona of an avuncular absent-minded professor, von Braun was ideal for the new medium. Attractive and charismatic, von Braun mixed intelligence and confidence to effectively convince Americans that manned spaceflight was a likely reality—so long as we shared the will and determination. And at that moment America was ascendant, a world leader in manufacturing and innovation with a burgeoning GDP and an expanding population, with many now earning college degrees. The possibilities for the future seemed limitless, despite the Cold War jitters that *Collier's* used initially to hook the attention of the public to buy their space series.

Perhaps no one better understood the collective American

24. Jim Korkis, "Man in Space," in USAToday.com, December 28, 2011.
25. Randy Liebermann, "The *Collier's* and Disney series," in *Blueprint for Space: Science Fiction to Science Fact*. Washington, D.C.: Smithsonian Institution Press, 1992.
26. "Man in Space" broadcast on *Disneyland*, March 9, 1955.
27. Ernst Stuhlinger, oral history interview by Mike Wright, December 17, 1992, Huntsville, Alabama. Cited in: Mike Wright, "The Disney-Von Braun Collaboration and Its Influence on Space Exploration," Marshall Space Flight Center History Office website, undated.

zeitgeist of the 1950s than Walt Disney. By developing family entertainment and amusement parks that appealed directly to national urges—a hunger and nostalgia for a simpler past and its values, a thirst for adventure, a dream to explore America's past frontiers—Disney was offering escape from the "Age of Anxiety." More than looking backward, Disney also understood the power of selling America an optimistic vision of its future: recycling the national myth of an open frontier in which space helmets replaced ten-gallon hats. Disney's insistence on scientific realism, with an emphasis on showing how mankind would eventually undertake a trip to the stars, made for revolutionary television. It was a reimagining of the American belief in limitless possibilities that divorced the concept of interplanetary travel from the juvenile trappings of melodramatic space operas.

In Washington, President Eisenhower viewed the "Man in Space" broadcast and the next morning called Disney to request a copy of the film. According to Ward Kimball, "the call came in to the studio and I guess at first the switchboard didn't believe it was the President of the United States."[28] The White House borrowed a copy of the film for two weeks, during which it was shown to Pentagon officials. A few months later, following Eisenhower's July 29th announcement that the United States would launch its first satellite in 1957, Senator Carl T. Curtis of Nebraska pointed out to his senatorial colleagues the great service that Walt Disney and his "Man in Space" program had performed for the government, and, by extension, the American people.[29] Yet when Disney attempted to capitalize on the developments in Washington by scheduling a third broadcast in September, von Braun wrote Kimball that he was "quite upset" with the publicity and feared that it would appear that he and Disney were attempting to take undue credit.[30]

Disney also received a request to borrow the film from two Russian space scientists, in the hope it might aid them in their efforts to gain support for the Soviet Union's rocket program. Their solicitation was summarily dismissed. Disney was already on record as a vocal anti-Communist, but according to Kimball, Disney's motivation for denying the request was more personal. Disney had a grudge against the Russian government, which, in the 1930s, had been loaned a print of *Snow White* on the understanding it would be returned within two

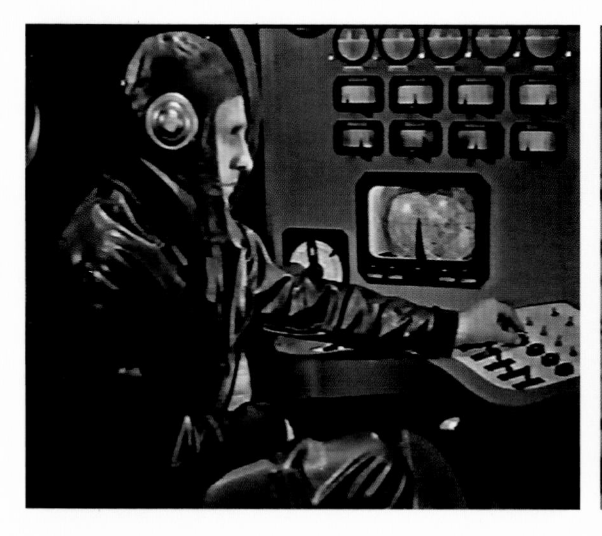

Selling Space to the Soviets: "Road to the Stars"

Two years after the broadcast of Walt Disney's "Man in Space" caused a sensation in America, the Soviet Union released their own speculative science documentary depicting man's future in the cosmos. *Road to the Stars* (*Дорога к звёздам / Doroga k zvezdam*) was a stunning, hour-long, color cinematic preview of man's first foray into space, including the construction of a huge revolving space station and a first landing on the Moon. A special effects triumph, the film was directed by filmmaker Pavel Klushantsev with technical assistance from Soviet rocketry pioneer Mikhail Tikhonravov. As *Road to the Stars* was nearing completion, the October 1957 launch of *Sputnik 1* forced the addition of a new ending celebrating Russia's orbital triumph. Americans got their first exclusive extended preview of scenes from Klushantsev's film on May 11, 1958, when Walter Cronkite introduced grainy black-and-white footage from *Road to the Stars* in an episode of the CBS News weekly documentary series *The Twentieth Century*. The episode, "Ceiling Unlimited," speculated on the Soviet Union's long-range plans in space in the wake of *Sputnik 1*, and included interviews with rocket scientists Wernher von Braun and Krafft Ehricke.

While conducting research in preparation for what would become *2001: A Space Odyssey*, filmmaker Stanley Kubrick obsessively tried to see every depiction of manned space flight previously committed to film, including productions made in the Soviet Union. It seems highly likely that Kubrick and co-author Arthur C. Clarke tracked down a rare print of *Road to the Stars* in the course of their initial planning in 1964, especially in light of a number of visual similarities between Klushantsev's film and Kubrick's epic. (MGM's first official announcement, in February 1965, reported that Kubrick's new film was to be called *Journey Beyond the Stars*.) For nearly fifty years, *Road to the Stars* was a cinematic footnote, more read about than seen. The Soviet film's recognition and rediscovery came as a result of a 2002 Danish documentary, *The Star Dreamer*, a reconsideration of Pavel Klushantsev's career.

Tomorrowland at Disneyland. Over on Main Street U. S. A., CAREFREE CORNER is the official Information Center maintained by INA and their independent agents

A new world of happiness

Ever see a family so thrilled! They're finding out all about space from the Space Man himself. And this is only one of Disneyland's exciting features. There are plenty more. Everywhere in Walt Disney's Magic Kingdom there is the gay, carefree sense of being lifted above worries.

This Disneyland 'magic of happiness' can be enjoyed all year 'round. The forward-looking people of the North America Companies (INA) have developed a new Family Accident & Sickness policy that helps families keep carefree. It's a *package policy* like the famous INA Homeowners and Tenants policies that add so much to the protection of homes. It is full-scale protection for a family and its members.

INA's new Family Security Program against accident and sickness is unique. It is really FOUR policies in ONE. And so flexible that you can make it your entire program or select just the portions needed to fill out the insurance you now have. Here is protection for the new world of happiness that healthy young families are building for themselves! Ask your INA agent about it.

* * *

Insurance Company of North America • Indemnity Insurance Company of North America • Philadelphia Fire and Marine Insurance Company • Life Insurance Company of North America • Philadelphia

INSURANCE BY NORTH AMERICA 

"Tomorrowland," Insurance Company of North America advertisement, 1957. *"Tomorrow can be a wonderful age. Our scientists today are opening the doors of the space age to achievements that will benefit our children and generations to come. The Tomorrowland attractions have been designed [with consultants Wernher von Braun, Willy Ley, and Heinz Haber] to give you an opportunity to participate in adventures that are a living blueprint of our future." So said Walt Disney at the dedication of Disneyland's Tomorrowland, in 1955. At once, Anaheim, California, became a national destination. Three things made this possible: promotion through Disney's outsized presence in the new medium of television, expanded commercial airline travel, and the building of the interstate highway system. Easy transcontinental travel did not allay ordinary concerns, however, but rather brought new ones. "What if we have an accident or get sick along the way?" INA, the Insurance Company of North America, developed a new series of policies to cover those contingencies, scoring yet another cross-promotional victory for both Disney and the space program. At INA, travel insurance related to Disneyland became a major campaign called "The Magic of Happiness," which included a variety of promotional giveaways: Disneyland postcards and a flexible phonograph record (manufactured by a company owned by Bing Crosby) featuring songs from Disney movies.*

28. Korkis, op. cit.
29. "On the Air with James Abbe," *Oakland Tribune*, August 14, 1955.
30. Korkis, op. cit.

NASA's "Other" Reading List

To promote the April 10, 1968, Southwest premiere of Stanley Kubrick's *2001: A Space Odyssey* at Houston's Windsor Cinerama Theater, MGM gave free passes to many NASA dignitaries, including all the astronaut flight crews. Some members of the aerospace community were as puzzled by the film's meaning, as were other audiences during *2001*'s initial release.

Astrophysicist Albert A. Jackson was a NASA employee in 1968 and attended one of the first screenings. Forty-three years later, he recalled how the film became a topic of discussion among members of the Apollo team at the Manned Spacecraft Center. "In the spring of 1968, I was 27 and was putting in very long hours as a lunar module simulation training instructor. It was intensive work, but wonderful fun; I saw each of the LM crews every day over the course of two years. When *2001* opened, I was working with, among others, Neil Armstrong and Buzz Aldrin, then members of the backup crew of Apollo 8, which was slated to be the second test of the lunar module and command module in medium Earth orbit. On the Monday morning following the film's premiere, a number of the guys were standing around the coffee pot talking about *2001* before heading over to the simulator. Many of them were trying to figure out what it all meant. I recall Buzz Aldrin picking up on the conversation and, without batting an eyelash, he explained the film's plot in detail—correctly, I should add—elucidating the ending of *2001* in terms of Arthur C. Clarke's other fiction, which he had read. He specifically cited a connection with Clarke's *Childhood's End* and its theme about the transcendent future of humanity. It was only one occasion of many in which Aldrin spoke about science fiction, and it was clear he read a lot during those years. He certainly wasn't an active fan, but he knew the literature."

Bill Anders, the lunar module pilot for Apollo 8, was also impressed. "I remember thinking it might be worth a chuckle to mention finding a monolith during our Apollo flight," Anders said shortly after the mission. Clarke's biographer, Neil McAleer, recounted that the author never forgave Anders for failing to follow through on this joke. "During the Apollo years, a lot of the employees at NASA read science fiction," Jackson recalled. "Many of the younger ones talked about watching the original *Star Trek* when it was airing on NBC. But I may have been the only true-blue fan working there who attended the World Science Fiction conventions. On one occasion, I remember recognizing a familiar face among a group of people receiving a tour of the Manned Spacecraft Center: it was illustrator Ed Emshwiller, whose work was often featured on covers of *Galaxy* and *Fantasy and Science Fiction* during the 1950s and 1960s, and had also earned secondary fame as a leading experimental filmmaker."

During the pre-production of 2001: A Space Odyssey, *director Stanley Kubrick and author Arthur C. Clarke hosted NASA visitors George Mueller, director of Manned Spacecraft Center, and astronaut Donald K. Slayton at MGM's Borehamwood Studio, in Hertfordshire, in September 1965. Left to right in the photo are: Frederick I. Ordway III, of NASA's Marshall Space Flight Center and special consultant on 2001; Slayton; Clarke; unidentified; Kubrick; and Mueller. After the tour, Mueller dubbed Kubrick's office complex "NASA East."*

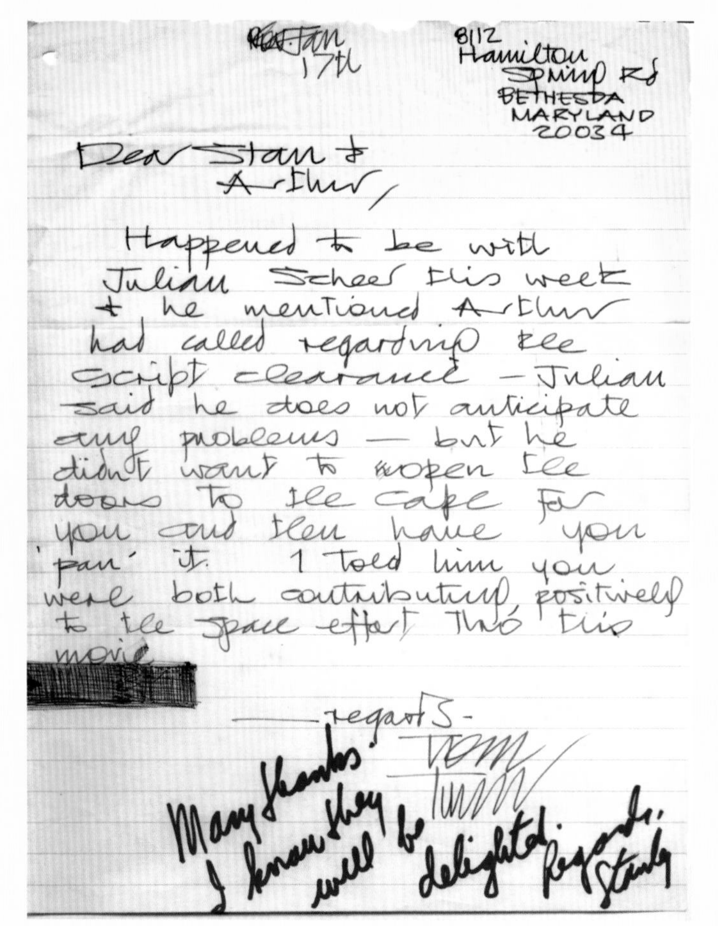

Jan 17th

8112 Hamilton Spring Rd
BETHESDA
MARYLAND
20034

Dear Stan & Arthur,

Happened to be with Julian Scheer this week & he mentioned Arthur had called regarding the script clearance – Julian said he does not anticipate any problems — but he didn't want to open the doors to the Cape for you and then have you 'pan' it. I told him you were both contributing positively to the space effort thro' this movie

— regards -
Tom
Turner

Many thanks! I know they will be delighted. Regards, Stanley

In planning 2001: A Space Odyssey, Kubrick engaged technical consultants to ensure the scientific accuracy of his project. Two Marshall Space Flight Center veterans, Frederick I. Ordway III and Harry Lange, were hired by Kubrick, at the suggestion of Arthur C. Clarke, in early 1965, and relocated to England for pre-production work that summer. As this letter shows, NASA's administrator for public affairs, Julian Scheer, was also consulted. In this letter, Thomas Turner, the president of the National Space Club, which involved representatives of industry, government, educational institutions, and private individuals, writes to Clarke and Kubrick referencing a conversation with Scheer about possible NASA clearance of the script. Turner mentions how he assured Scheer that the planned film would "contribute positively to the space effort." Amended is a note in Kubrick's hand predicting that NASA "will be delighted" with the film. The date of this letter appears to be January 1966, three weeks after principle photography on the film commenced. (Document furnished by Thomas Turner.)

weeks. After holding the print for more than a decade, the Russians finally returned the errant *Snow White*, only to have the studio discover that the print was defaced with Russian subtitles and badly scratched from countless—unremunerated—showings.[31]

Following three national broadcasts, "Man in Space" was later coupled with Walt Disney's second major television phenomenon of 1955, when it appeared in abbreviated form as a documentary short before theatrical screenings of *Davy Crockett and the River Pirates,* a compilation of two *Disneyland* episodes featuring Fess Parker as the legendary folk hero. It was fitting that, at the dawn of the space age, the high frontier of tomorrow shared the screen with the mythic frontier of American past.

Meanwhile, in Anaheim, Disneyland's "Tomorrowland" opened, featuring as its central attraction the TWA Moonliner, an eighty-foot, needle-nosed rocket ship designed by Disney's John Hench with advice from von Braun. Branded with the TWA logo and the distinctive red-and-white color scheme familiar from the airline's contemporary fleet of Lockheed Constellations, the Moonliner was intended to depict a realistic idea of a commercial spaceliner of 1986. It was also an early form of partnered corporate product placement, for which TWA provided sponsorship. Ironically, when an American plastic model kit of the Moonliner was issued in the United Kingdom, the English manufacturer Selcol dropped Walt Disney's name, choosing instead to market it as "Dr. Wernher von Braun's Moonliner," a dramatic indication of how rapidly the man who designed the V-2s that hit London in 1944 and 1945 had—with the aid of the American intelligence services—successfully accomplished an amazing feat of personal rebranding.

Psychologically, Disney and von Braun shared a number of distinct personality traits, the foremost being an unwavering dedication to seeing their personal visions made concrete. Self-confident, optimistic, and gifted with an ability to inspire others to do their best in pursuit of their goals, each had an uncanny, instinctual skill for marketing and persuasion that came to shape American life during the mid-20th century. One was often described as politically naïve, while the other sometimes labeled a political opportunist, but for both men, politics mattered only insofar as it aided the realization of their dreams.

Ten years after "Man in Space" aired, while Project Apollo was underway, von Braun invited Walt Disney and others involved in the "Tomorrowland" films to tour the Marshall Space Flight Center in Huntsville, Alabama. At the time von Braun expressed his hope that this might rekindle Disney's enthusiasm in the subject and spark a new movie about manned space flight.[32] However, despite a Disney interview with a newspaper following the tour, in which he said, "If I can help through my TV shows . . . to wake people up to the fact that we've got to keep exploring, I'll do it," no film resulted.[33] The following year Disney was dead, never able to witness the sight of a man walking on another world, a feat he indirectly helped set in motion.

When the launch of *Sputnik 1* shocked the world, marking the dawn of the space age, the United States was far better equipped for the challenge than it realized. And Walt Disney, *Collier's*, and Wernher von Braun played pivotal roles in that preparation by envisioning an optimistic future: a reconciliation of the romantic fantasies of the past with the practical realities of tomorrow. In less than a decade, space travel had emerged from the realm of children's adventure stories and the domain of rocketry and science fiction hobbyists to the world of front-page headlines. And just as Verne had imagined nearly one hundred years earlier, an unprecedented public-private partnership between government, contractors, and the media would launch a three-manned crew from Florida destined for the Moon. ⦿

Using Wernher von Braun's name in advertising was not considered a disadvantage when a British toy manufacturer marketed plastic model kits of the Disney Moonliner in the United Kingdom in the late 1950s.

31. Ibid.
32. Wernher von Braun, unpublished handwritten note to Bart Slattery. Dated "3/6," presumed to be March 6, 1965.
33. Bob Ward, "Walt Disney Makes Pledge to Aid Space," *The Huntsville Times*, April 13, 1965.

NASA Brand Journalism: Finding a Voice

A YEAR AFTER the Soviet Union's surprise launching of *Sputnik 1*, the National Aeronautics and Space Administration formally opened for business on October 1, 1958. Its official seal, with its streamlined red delta and symbolic representations of the Earth, Moon, stars, and an orbiting satellite, looked like nothing previously seen in Washington; it was confident, visionary, and futuristic. And in keeping with that new brand, NASA conceived a truly futuristic approach to public affairs during the course of its formative year.

"In servicing the press, the PIO [Public Information Office, later the Public Affairs Office] seeks to function as a precision-ground mirror, faithfully reflecting the activities of NASA," wrote Walter T. Bonney, the head of NASA's nascent PIO, in a 1959 policy memo to NASA administrator Dr. T. Keith Glennan. While Bonney would not serve a long time in this role, his memo outlined a philosophy adopted by most of the agency's public affairs officers during the entire Apollo program, and served as a template for how the office interacted with the public and the press.[1]

In the lengthy memo, Bonney laid out a vision for NASA's approach to PR, and in so doing defined a technique that later practitioners would refer to as "brand journalism" and "content marketing"—taking a journalistic approach to tell a story by clearly reporting the facts, rather than pushing an agenda or selling a message in the manner of traditional marketers or publicists. He wrote: "In practice, PIO staff work as reporters within the agency, seeking out newsworthy information from NASA technical personnel and processing it into a form useful to the press. The press uses this product of the PIO—the releases, the pictures—much as it uses the product of the wire services, with this important difference: It rewrites the product of the PIO and in the doing, makes the product its own."[2]

To support his vision of a NASA public affairs group operating not as pitchmen but as reporters, Bonney, a former journalist, actively sought and recruited staff with print and broadcast media experience. Among those Bonney enlisted during the early days of the agency were Paul Haney and Jack King, who would soon play key roles during the Apollo program. "The core contingent of NASA Public Affairs people—just about all of us—were ex-newsmen," recalled King. "We were good writers, and we knew the news business. That made a major difference in the whole operation."[3]

The NASA Public Affairs Office's growing ranks of journalists understood what constituted a good story and what details appealed to the press. Thus, NASA created materials that addressed reporters' needs in press releases, bylined articles, background materials, and in sponsored media symposiums, television newsreels, and fully produced radio broadcasts complete with interviews and sound effects.

NASA's brand journalism approach also created a unique relationship with the nation's press that very few corporations or government agencies have ever enjoyed. Brian Duff, former Director of Public Affairs at the Manned Spacecraft Center in Houston during Apollo 11, and later head of NASA Public Affairs, contends that "the relationship between NASA and the media was a pioneering effort, in terms of cooperation and mutual trust, and, in more than twenty-five years, there were very few breaches of that trust on either side. Anyone who thinks that NASA had an adversarial relationship with the media just doesn't know what a real adversarial relationship is. The Nixon White House had a truly adversarial relationship with the media. The situation at NASA was nothing like that. NASA and the press had a sweetheart relationship."[4]

Publicity vs. Public Information

Because of this special relationship with the press, Bonney's approach to brand journalism was not without controversy. No single public affairs decision generated more strain and elicit-

OPPOSITE: NASA Public Affairs staff shown working in the Mission Control Center in Houston on June 7, 1965, during recovery operations of the Gemini 4 mission. From left to right: Bob Hart, Paul Haney, Al Alibrando, and Terry White.

1. Ginger Rudeseal Carter, "Public Relations Enters The Space Age: Walter S. Bonney And The Early Days of NASA PR," paper presented at the 1997 Association for Education in Journalism and Mass Communication (AEJMC) Chicago Convention
2. Ibid.
3. Jack King, interview with the authors, November 9, 2011.
4. Brian Duff, Smithsonian National Air & Space Museum Oral History Interview, May 1, 1989.

Walter T. Bonney (right), the head of NASA's nascent Office of Public Information, and Robert R. Gilruth (seated), head of the Space Task Group and later the first head of NASA's Manned Spacecraft Center.

ed more criticism than the exclusive *Life*/World Book Science Services contract. Executed in August of 1959, the contract granted *Life* magazine exclusive access to the personal stories of the astronauts and their families. In exchange, the original Mercury astronauts divided compensation that equaled roughly $25,000 annually for each of the seven, as well as individual life insurance policies of $100,000.[5] As more astronauts entered service, the compensation level for each astronaut diminished.

Some newspaper columnists took NASA to task, implying the astronauts were "cashing in" on experiences paid for by taxpayers. The situation only got messier after the media learned of an ill-advised attempt by the astronauts to invest the compensation from their contract in a Cape Canaveral-area hotel. When stories surfaced that the astronauts had accepted offers of free homes from a Houston real estate developer, the White House intervened.

Bonney saw the *Life* arrangement differently. In his view, the contract served to separate the personal lives of the astronauts, who would inevitably become celebrities, from NASA's public affairs duty to provide information to the press and the public. "The distinction between publicity and public information must be kept constantly in mind," Bonney wrote in his memo in 1959. "Publicity to manipulate or 'sell' facts or images of a product, activity, viewpoint, or personality to create a favorable public impression, has no place in the NASA program." Instead, Bonney believed, and instilled in his staff, the understanding that their job was to "furnish Congress and the media with facts—unvarnished facts—about the progress of NASA programs."[6]

Despite Bonney's attempt to reconcile his approach with the *Life* contract, NASA exerted considerable control over the images and stories that appeared in the magazine. As part of their involvement, both the NASA Public Affairs Office and the astronauts reviewed copy prior to publication. From a practical standpoint, the *Life* contract also helped NASA deal with the burgeoning global media interest in the lives of the astronauts' families. "The original astronauts wanted to protect their lives and those of their families," recalled Gene Cernan, commander of Apollo 17 and one of the third group of astronauts, selected in 1963. "I mean, the press early on was following our kids to school. The exclusivity of the *Life* arrangement inhibited people from peeking over our fences and that kind of thing."[7]

"The contract got approved in order to keep us from being inundated," added Walt Cunningham, a fellow member of the third astronaut group, who flew on Apollo 7. "It protected us from being distracted." And for the astronauts and their families, the $100,000 life insurance policy within the contract was important. "It's always referred to simply as the *Life* contract, but it was the '*Life* and World Book Science Services' contract. In fact, World Book and their parent company, Field Enterprises, carried the bigger load, something like 60%. World Book Science Services came in and paid several hundred thousand dollars to each of the families of the Apollo 1 fire."[8]

"It was also a PR leveling factor, and I think that is pretty important, too" added Cernan. "Imagine if there had been no contract, and some of the guys shared their personal stories, and wanted to be big-time heroes and have their picture on the front page of the papers. Now imagine the other guys who did not, who wanted to keep their personal lives private. People would hound them, follow their wives, follow them to the mall—that's what the contract tended to prevent. Once the contract was established, the entire press corps realized what they were going to get access to, and what they were not."

President Kennedy Intervenes

When the contract came up for renewal in 1962, an editorial in the *New York Times* declared that the United States government should not allow the astronauts to reap "enormous private profits" from participating in "a great national effort." The

5. Harlen Makemson, *Media, NASA and America's Quest for the Moon* (New York: Peter Lang Publishing, Inc. 2009), p. 49.
6. Ibid., p. 91.
7. Gene Cernan, interview with the authors, February 24, 2012.
8. Walter Cunningham, interview with the authors, February 14, 2012.

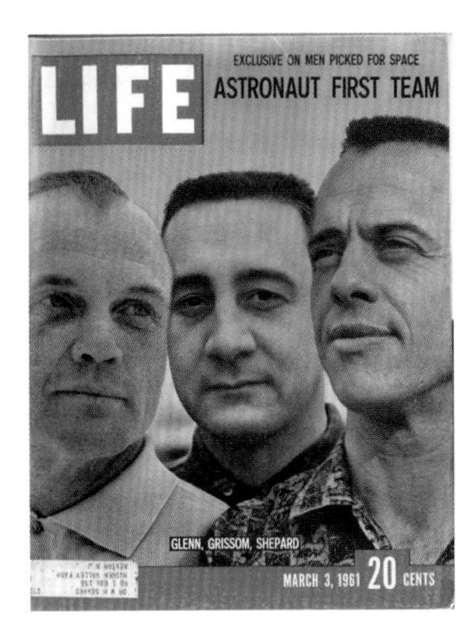

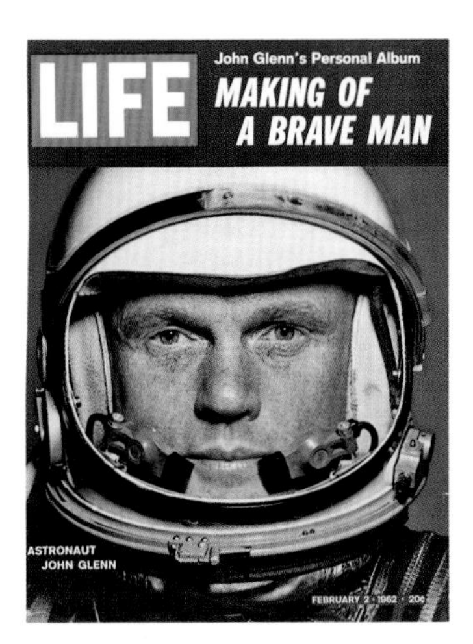

LIFE *magazine. On August 5, 1959, the original Mercury 7 astronauts signed an exclusive, $500,000 contract with Life/World Book Science Services, which allowed the publication exclusive access to the astronauts' personal stories – including access to their home life, wives, and children. The astronauts and NASA believed this exclusive contract would allow for the astronauts to have better control over their private lives, and would wall-off their families from the immense media intrusion. In addition, the money would help augment the astronauts' relatively meager military salaries and compensate them for the worldwide scrutiny and attention they were receiving. The astronauts and NASA worked with Life/World Book Science Services to carefully craft the image of the astronauts, not as military men, but as middle-class average family men thrust into service for the good of their country. While functionally efficient for NASA and the astronauts, the arrangement proved to be controversial, and it angered other news outlets, who were denied direct access to the lucrative, exclusive stories that were printed not only in* Life *magazine but also as first-person account spin-off books such as* We Seven. *When the contract was up for renewal, only a personal plea from John Glenn to President Kennedy saved the arrangement. It was later expanded to include World Book Enterprises, and the fees began to be split up by a growing number of astronauts in the program. As an additional incentive, Life/World Book Science Services also provided the astronauts with important life insurance coverage, which was otherwise denied them under government service regulations.*

Times concluded that the government contract followed an inappropriate policy when it allowed public-domain astronaut stories to become a "private payload." Understandably, the era's two leading news magazines had contrasting views of the contract. *Life*'s sister publication, *Time*, did not seriously criticize the contract. Its chief competitor, *Newsweek*, did. According to *Newsweek*, the contract was an "embarrassing financial arrangement" with contract negotiations "more appropriate to the film *Cleopatra* than to a "serious scientific endeavor."[9]

The negative publicity and attention to the *Life* contract was not lost on NASA management or on President Kennedy, and there was general agreement in Washington that the contract should not be renewed. But the Mercury astronauts banded together and took their case to Vice President Lyndon Johnson.[10] Soon after, John Glenn addressed the matter personally with President Kennedy.

During the early summer of 1962, Glenn and his family were invited to visit the president in Hyannis Port, Massachusetts. Glenn accepted the offer and later reported that he and his family had "a wonderful time." At the end of a day filled with water-skiing and swimming, Glenn and the president relaxed on the fantail of the *Honey Fitz*, Kennedy's yacht. As Glenn later recalled in a 1964 oral history,[11] it was at that moment that Kennedy asked him directly about the *Life* contract, and solicited Glenn's thoughts on the subject:

> So I went into some of the details of why we had the contract to begin with and what the contract actually consisted of, which was generally misunderstood at the time, and still is, as far as that goes. But we went into some detail of how the contract did not cover experiences from the space flight, that the contract covered permitting the *Life* people to talk to our families, our children, and us—in other words, it gave them access to our homes. How do these people live and what was their childhood background? What do they think? What do they do when they go to church? What do they do when they play with the kids? What do they do when they go to the swimming pool? In other words, the little personal day-to-day family life and activity was the only thing that was for sale on this contract and that's the only thing that has remained for sale on this. It's just been the personal background story, not the story of the flight. Well, he hadn't understood this in this vein at all, and I explained this very carefully, and that it was the only thing that was for sale—that I was as much as anyone against selling what we were actually doing on the flights themselves, and if anyone could show me anywhere we ever held back one iota of information that we gained on a flight to give to any particular person, I'd like to know about it. That information has been open to everyone on just as full a measure as it has been to *Life*. I'd be happy to give back any money I had received if anyone could show me where I had withheld anything from other press conferences or interviews. . . . We discussed this for probably twenty minutes to half an hour. We went into all phases of it: how this had worked out during my flight; how through this we had been able to control some of the press activities with the family because we did have this contract. Of course, to put it on a crass commercial basis, this thing had been very good in that it guaranteed that I could give my children the education that I wanted to give them. So from a strictly personal standpoint, I was all for it and thought it had worked to our advantage and to NASA's

9. Kristen Amanda Starr, "NASA'S Hidden Power: NACA/NASA Public Relations And The Cold War, 1945–1967." Ph.D. dissertation, Auburn University, 2008. pp. 268–269.
10. Makemson, op. cit., p. 94.
11. John Glenn, John F. Kennedy Library Oral History Project Interview, June 12, 1964, p. 8.

LEFT: An example of NASA brand journalism, utilizing a reprinted article from the New York Times *by science writer Walter Sullivan, titled "What Earthly Use is the Moon?"*

RIGHT: A Kennedy Space Center PAO publication from 1968, including a NASA-produced advertisement promoting the low cost/high benefit of space exploration on a national per-capita basis of just two cents.

WALTER SULLIVAN is science editor of The New York Times. • © 1966 By the New York Times Co. • Reprinted with permission by the National Aeronautics and Space Administration.

"In the old days, when getting away meant a trip to Florida, I could never bring myself to try water skiing, and I'm certainly not going to try *this!*"

The speaker was an elderly tourist watching young men with huge plastic wings strapped to their arms. By flapping them vigorously they were able to fly up toward the lofty roof of white fabric covering the community like a miniature sky, then swoop down at peculiarly slow speed. It was slow because this was a colony on the moon, where gravity is only one-sixth as strong as that on earth, a fact that enabled the local residents to develop muscle-powered flight as a sport within the air-inflated, air-supported domes that covered the lunar colonies.

Such is one of the more fanciful pictures of life on the moon that have been painted by those looking into the future. Muscle-powered flight, in the manner of that mythical father-and-son team Daedalus and Icarus, would require much practice—probably at the expense of many black-and-blue spots. But, as noted by the inventor of the hypothetical sport, Arthur C. Clarke, there are no physical 'aws or limits to our engineering capability that stand in the way. In a recent book he proposed that the moon will have its native inhabitants—men and women who live out their lives on the moon, helping to exploit its resources and man its observatories.

Clarke, who dreams of the future with a good deal of scientific sophistication, also writes science fiction. He represents the far-out school of thought, but the dramatic manner in which earlier dreams of this sort, such as those of Jules Verne, have been fulfilled reminds us that what seems far-fetched today may prove the reality of the not-too-distant future.

Why should anyone bother putting colonies on the moon? Does the moon have anything the earth does not? This year's report of the Soviet Union on its space program points out that the environment of the moon is so different from that of the earth that minerals should be found there unknown to students of this planet's rocks. The moon has virtually no atmosphere: in fact, its vacuum is more nearly complete than any achieveable in ordinary laboratories. Its surface materials have therefore not been oxydized in the manner of earth rocks. They have been subjected to millions—or even billions—of years of intense radiation from the sun, unshielded by any air blanket. All of these factors must have produced substances with properties foreign to those of our landscape.

The moon may thus open new vistas in chemistry and mineralogy. But the uses of the substances to be discovered there cannot be guessed until their nature is known. One can only look into the past and recall that the opening of every new realm of scientific knowledge has brought with it revolutionary changes in our capabilities.

Actually, in our museums, there may be hints of what the astronauts will bring back from the moon. There are certain of the stony meteorites whose composition differs from anything found on earth. They are packed with tiny objects, called chondrules, that often look like grains of rice.

It has been proposed that these meteorites may be chips knocked off the moon by the impact of great iron meteorites. The irons are thought to be fragments from the core of one or more asteroids shattered by some collision in the distant past. There are thousands of asteroids circling the sun between the orbits of Mars and Jupiter—a region known as the asteroid belt. These objects never come near the

LUNAR INDUSTRY—The moon's rocks may offer man new raw materials. Here, a visualization of a chemical plant for making rocket fuel on the moon.

General Electric art work by Roy Scarfo

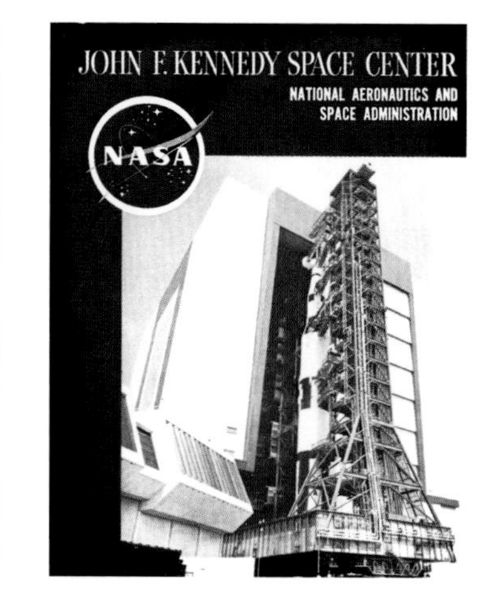

> and the government's in the way we had handled it; and, if we could just get across to people the idea that this was not a sale of our experiences in flight, I thought this would be generally accepted."

Paul Haney, soon to become Director of Public Affairs at the Manned Spacecraft Center, agreed with Glenn, and supported Bonney and the contract, as did Project Mercury-era public affairs officer Col. John "Shorty" Powers. "Major newspapers editorialized about how these new icons 'with feet of clay' had 'sold out' to journalistic picture/news giant *Life* magazine," Haney recalled in an essay titled "Spinning Space in the Cold War." "We at NASA tried to make the point that *Life* had bought the personal story of the astronauts. The technical part of their job . . . would somehow be reported fully, but none of us knew precisely how at the time."

Glenn's direct appeal to Kennedy was decisive and the contract was renewed, though not before a special study group within NASA reviewed the situation and made a number of recommendations. As a result, in September 1962, NASA announced a new policy aimed at giving all news media equal access to the astronauts' stories about their missions and training. Specifically, NASA added a second, post-flight news conference in which a smaller, select group of journalists were given an opportunity to interview the astronauts in a more relaxed setting. No longer could a publication advertise or imply that it featured the astronauts' exclusive stories about their mission. And no longer could the astronauts publish or collaborate on a published piece without prior approval.[12]

Additional caveats and clauses were added, significantly guidelines to prevent the astronauts from making any investments that hinted of impropriety, including commercial endorsements. The *Life*/World Book contract was renewed with each new class of astronauts, and continued throughout the Gemini and Apollo programs, though its impact—both financially and with the American public—diminished greatly as the 1960s progressed. Not only did spaceflight become less of a novelty, but as the NASA astronaut corps grew in size to number more than 50 members by 1966, it became increasingly difficult to portray each astronaut as a distinct personality.[13]

Still, this singular example of NASA brand journalism was instrumental in fostering an indelibly positive image of the astronauts. Nearly four decades later, Haney reflected, "*Life* over the period of two years introduced the all-American

12. NASA News Releases, September 16, 1962.
13. Starr, op. cit., pp. 269–270.
14. Paul Haney, "Spinning Space in the Cold War," cited in *The Harvard International Journal of Press/Politics*; June 1998, 3:126–131.
15. "Haney, Voice of Apollo, Quits NASA." The Associated Press, April 28, 1969.

flyboys to the world. Backyard photos of families barbecuing. In depth portrayals by several of *Life*'s better writers. All of which became a book called *We Seven*. There is no way NASA could have gotten the pilots off to a better public start."[14]

Be that as it may, it was not beyond NASA to leverage the work of noted science and technology writers to help tell their story. In 1966, the *New York Times* published a feature "What Earthly Use Is the Moon?" by science editor Walter Sullivan, which promoted the Moon as a potential scientific observation base and a new source of raw materials. Not long after, NASA published it as a reprint for distribution to the press and public to help bolster the Apollo Program. The artwork was by General Electric technical artist Roy Scarfo.

From its earliest days, NASA's Public Affairs Office employed experienced journalists and writers to create its press materials. Walter T. Bonney believed the agency's message was conveyed most powerfully when newspaper editors and reporters around the country used NASA's materials word-for-word without attribution, thereby conveying the publication's endorsement of the content. In addition to printed and filmed pieces distributed as "ready made" news material for the press, NASA Public Affairs also targeted radio stations with prepackaged, fully produced interview programs of 14½ minutes duration, distributed on vinyl records and open-reel audiotapes. These "NASA Special Reports" often included exclusive interviews with astronaut crews, NASA management, and lunar scientists, thereby giving smaller stations with limited news budgets an opportunity to offer their audiences interesting coverage of the space program that otherwise would have been unavailable to them.

Finding NASA's Voice in 1969

"We don't put out publicity releases. We put out news releases." —Julian Scheer, NASA Public Affairs chief

"When you take on a $4 billion bureaucracy, you have to be prepared to lose your job," Paul Haney, said upon angrily resigning his post as head of Public Affairs at the Manned Spacecraft Center in Houston, ten weeks before Apollo 11 would touch down on the moon.[15]

Not satisfied with merely quitting over what he described as a "badgering series of exchanges" with Julian Scheer, NASA's

NASA's Films

During the height of the manned space program of the 1960s and 1970s, NASA produced and distributed scores of documentary films intended for the general public. While sometimes unfairly classified as "industrial films," NASA's documentaries garnered a number of awards and were notable for their content and objectivity. Coincident with the early American manned space program, most American public schools started introducing the use of audio-visual equipment in the classroom, notably 16mm sound film projectors.

Each of the seven regional NASA offices maintained extensive film libraries and had a desk that coordinated the bookings, dispatching 16mm sound film prints to secondary schools, churches, Boy Scout troops, libraries, rocketry clubs, and other non-profit organizations. (NASA loaned the films free of charge on condition there would be no admission fee; return postage was the only expense incurred.)

NASA's documentary films were also distributed to commercial network affiliate television stations, thus extending NASA's public relations exposure. It was not unusual for the stations to program NASA films during early morning or late night timeslots, a part of the daily schedule that seldom attracted viewers or earned advertising income. These thirty-minute documentaries covered a range of subjects, from notable manned missions and evolving theories of cosmology, to a fictional story intended to illustrate how satellites dramatically changed the lives of average Americans. NASA's thirty-minute film about the Apollo 13 mishap, *"Houston, We've Got a Problem,"* was distributed only a few weeks after *Odyssey*'s triumphant return, and concentrated on the crisis team working at Mission Control. The editing, music tracking, and production was left to an outside contractor, in this case A-V Films of Houston, Texas, which assembled the documentary from exclusive NASA footage shot at the Manned Spacecraft Center, and news film from independent commercial sources.

Throughout the '60s and '70s, many television stations regularly aired *Aeronautics and Space Report*, NASA's periodic newsreel-like compendium of recent achievements and progress reports. A typical installment might cover a recent planetary probe, a weather satellite launching, an investigation into airline safety, or heat-shield research. Most installments were less than five minutes in length, making them popular with weekend daytime schedulers at network affiliate stations, who often needed short duration content to fill air time, such as when a baseball game ended earlier than expected.

LXPO42201-4/22/69-SPACE CENTER,HOUSTON:Paul Haney,the world-famed"Voice of Apollo",said 4/22 he has been relieved of his duties as chief spokesman for the Manned Spacecraft Center and ordered to report to a new space agency job in Washington 4/28.Haney,holding achievement awards he received for his work on Apollo 8 and 9,said he is considering appealing through the Civil Service Commission the order relieving him of the post. UPI TELEPHOTO waf
NASA
MANNED SPACECRAFT C
Office of Public Affairs

UPI wire photo of Paul Haney, shortly after he was reassigned to Washington, three months before the Apollo 11 mission. He is seen here holding some of his achievement awards as he embarks on an ill-fated PR offensive.

head of Public Affairs in Washington, Haney went on the offensive with reporters by expressing shock and dismay at his treatment by NASA management, both in Houston and in Washington. After being informed by Scheer that he was being reassigned to Washington, Haney walked out of his offices at the MSC, went over to Ed Delong's office at United Press International, and called a "semi-official, semi-private" press conference.[16]

Nervously rubbing one hand across his head, Haney announced with a sad smile that he had just been relieved of his duties.[17] He was stunned to learn he was being reassigned, following what he termed "weeks of harassment" by Scheer, and illustrated the tensions by relating an exchange in which Scheer accused him of being "a goddamn liar."[18]

Haney, who had been hired by Walter Bonney, had maintained close relationships with many in the news media.[19] So it was not surprising that a number of reporters that Haney had met during his long career at NASA in Washington and Houston rallied around him, writing stories that were quickly picked up by hundreds of newspapers across the country. Almost all of the coverage was sympathetic to Haney, and somewhat incredulous over the timing of his departure.

Popular syndicated writer Bob Considine underscored the ill-timed nature of the move, suggesting it would be like the New York Yankees dumping famed play-by-play broadcaster Red Barber on the eve of a pennant race.[20] Reflecting on the incident, Haney's replacement, Brian Duff, said, "When you replace the Public Affairs officer at Houston ten weeks before the first landing, that's a decision that has to be made with some consideration."[21]

"We did not shake hands when we parted," a defiant Haney told the press in Washington after a marathon three-hour meeting, in which Haney reported turning down reassignment and formally tendering his resignation. "We found no area of accommodation at all."[22]

The press, gearing up for the rapid succession of launches of both Apollo 10 and 11, characterized the exchanges between Haney and Scheer as a "bitter battle," and quickly compared the action with that of Scheer's 1963 reassignment of MSC Public Affairs head Shorty Powers, a highly publicized decision that resulted in Haney taking over Powers's responsibilities. In an ironic case of *déjà vu*, Haney saw the move as nothing short of retribution for his positive PR accomplishments; he suspected he had been viewed by headquarters as "hogging the news spotlight."[23] Haney further suggested that Scheer was "jealous of or unhappy about [Haney's] fame."[24]

Powers, never one to shy away from the Houston media, went on the record to exact his own pound of flesh: "I think Paul Haney was fired, and fired in the same fashion that I was, by the same guy," he told United Press International. Powers, came to NASA Public Affairs "on loan" from the Air Force to provide direct PR support to the original Mercury 7 astronauts, and to be the first head of Public Affairs at the newly constructed MSC. Powers had a flair for self-promotion and a bit of a temper, and his style openly fell into conflict with headquarters.

"In the early days you thought you were hearing about space, but you really were hearing Shorty Powers telling you about space," recalled Duff. Powers was the first "Voice of the Astronauts," and he, like Haney, became famous for his launch and mission commentaries—even when he was making up details on the spot for effect. "Shorty invented such words as 'A-OK' and a phrase that those of us in the business used: 'We have a calm, cool, and collected astronaut.' He didn't have any more idea that it was a calm, cool, and collected astronaut than some guy in the press box; it was totally subjective on Shorty's part. But he thought his job was to inject color into the program. Later, we got away from that sort of thing and tried very

16. Edward K. Delong, "Muted Haney Sore over Job Shift," United Press International, April 23, 1969.
17. Ibid.
18. "Haney Relieved of NASA Duties." The Associated Press, April 23, 1969.
19. Paul Haney, Johnson Space Center Oral History Interview.
20. Bob Considine, *On The Line: People . . . Places*, May 1969.
21. Brian Duff, Smithsonian Air and Space Museum Oral History Interview, 1989.
22. "Haney, Voice of Apollo Quits NASA," op. cit.
23. Paul Haney, "Spinning Space in the Cold War," op. cit.
24. "Paul Haney Ordered to D.C.: 'Voice of Apollo' Blasts Off Against Job Switch." United Press International, April 24, 1969.

hard not to inject color, and let the voices speak for themselves," Duff remembered.

Powers wasn't a favorite of the newsmen covering Mercury, particularly CBS News's Walter Cronkite. "The guy who was the voice of Mercury control during those early space flights became so cocky that he designated himself as the eighth astronaut. He got to be something of a butt of a lot of jokes and a laughingstock in a sense, around the space program even among the astronauts themselves."[25] Cronkite accused Powers of negligence during Scott Carpenter's *Aurora 7* flight in May 1962. Powers failed to inform the press that, when voice communications were lost with *Aurora 7*, NASA was still receiving relevant and up-to-date biomedical information, thereby assuring the flight controllers that Carpenter was fine. Cronkite says he was forced to go on the air and fill 40 minutes of live television, describing the fate of the mission as being in doubt, even though NASA knew Carpenter was in no danger. Cronkite believed that Powers was never properly chastised.[26]

It was Powers's showmanship, his independence from NASA's main office in Washington when handling public affairs matters in Houston, and his tendency to make up details rather than just reporting the facts, that resulted in his reassignment to Washington. But his fatal error was in favoring television reporters over print journalists, to the point of handing out mission flight plans to network correspondents, but failing to do the same to the print media's counterparts. Julian Scheer would not tolerate any sign of bias or favoritism, or any challenge to the open-program concept of equal and fair distribution of information. Powers "really saw the space program as a video, audio, sort of thing, but I saw it in a broader context," said Scheer, in an interview years later.[27]

"The sequence of events in Paul [Haney]'s case was much the same as mine," Powers would claim to the press, at the time of Haney's departure.[28] It should be said, though, that Powers's departure from Houston was accomplished in a more professional manner. Powers never called a press conference to publicly air his professional grievances, and he accepted the reassignment to Washington. In return, NASA officials issued extremely positive press statements about his contributions. NASA Administrator James Webb highlighted the fact that Powers had played an important part in the success of the Mercury program, and that "all of us in NASA appreciate his

Col. John "Shorty" Powers, the first head of Public Affairs at MSC in Houston, and on-loan to NASA from the military, was considered the "Voice of the Astronauts," and who considered himself, according to Walter Cronkite, the "8th Astronaut," is seen here with Col. John Glenn shortly after Glenn's historic orbital flight in 1962.

contributions."[29] MSC Director Robert Gilruth stated he had "great admiration for the excellent job Colonel Powers has done for the manned spaceflight program and for the country."

Scheer, for his part, publicly took the high road, and refused to comment on the details. In his appointment of Brian Duff, a veteran newsman, to replace Haney, he simply stated that he thought reassignment of Haney "would make the best use of Paul's talents."[30] For his part, Haney said the move "hurt like hell."

Out of Sync

After the initial press conference, Haney continued to talk to the press leading up to the launch of Apollo 11, causing a series of minor scandals by discussing private, internal issues that generated considerable bad press for NASA. The most egregious instance involved interviews he gave to the Associated Press, asserting that Neil Armstrong had "pulled rank" over Buzz Aldrin to become the first man to walk on the moon.[31] The widely published story created a flurry of negative headlines and speculation of infighting among the crew just days before launch, pressuring Deke Slayton and NASA administrator George Low to issue statements denying Haney's claims.[32]

Haney also cast aspersions about what he viewed as the "lucrative" *Life* contract, even though he had supported it earlier. In a syndicated article by *New York Times* writer Richard

25. Walter Cronkite and Don Carleton, *Conversations with Cronkite*, 2010; Austin: University Press of Texas, p. 234.
26. Ibid.
27. Quoted in Ginger Rudeseal-Carter, op. cit.
28. "Fired Same Way, Says Powers," United Press International, April 24, 1969.
29. "Haney Succeeds Powers as PAO," in *NASA MSC Space NEWS Roundup*, vol. 2, no. 21: August 7, 1963.
30. Edward K. Delong, op. cit.
31. "First Man on the Moon? Rank has its privileges," Associated Press, June 27, 1969.
32. "Pigtail Bonny Ready for Ride," Associated Press, June 18, 1969.

Memorandum from Paul Haney, Chief of Public Affairs, Manned Spacecraft Center, Houston, with annotations by Alan B. Shepard, Jr., Chief, Astronaut Office. *Paul Haney was responsible for the dramatic shift toward live mission reporting as a result of a direct appeal to President Kennedy during the Mercury program. Prior to that, all public information releases were on a time-delayed basis, and all flights were announced only after "fire in the tail," that is, after the rocket had left the launch pad.*

In addition to his reporting duties, Haney supported the policy mandate to make public all mission transcripts, to which some astronauts objected, in the belief that raw, unedited language might tarnish their public image. The annotations here are by astronaut Alan B. Shepard, the first American to go into space, and who, as chief of the Astronaut Office, governed the astronauts' access to the press. Haney's annotations in the upper right-hand corner refer to NASA HQ policy on treatment of astronaut personal information, including the confidential nature of their Personal Preference Kits, and their right to privacy and image protection in case of a tragic event.

The battle between the Public Affairs Office and the Astronaut Office for access and information was an ongoing source of tension throughout the duration of the program, as were policy conflicts between Washington and the field offices.

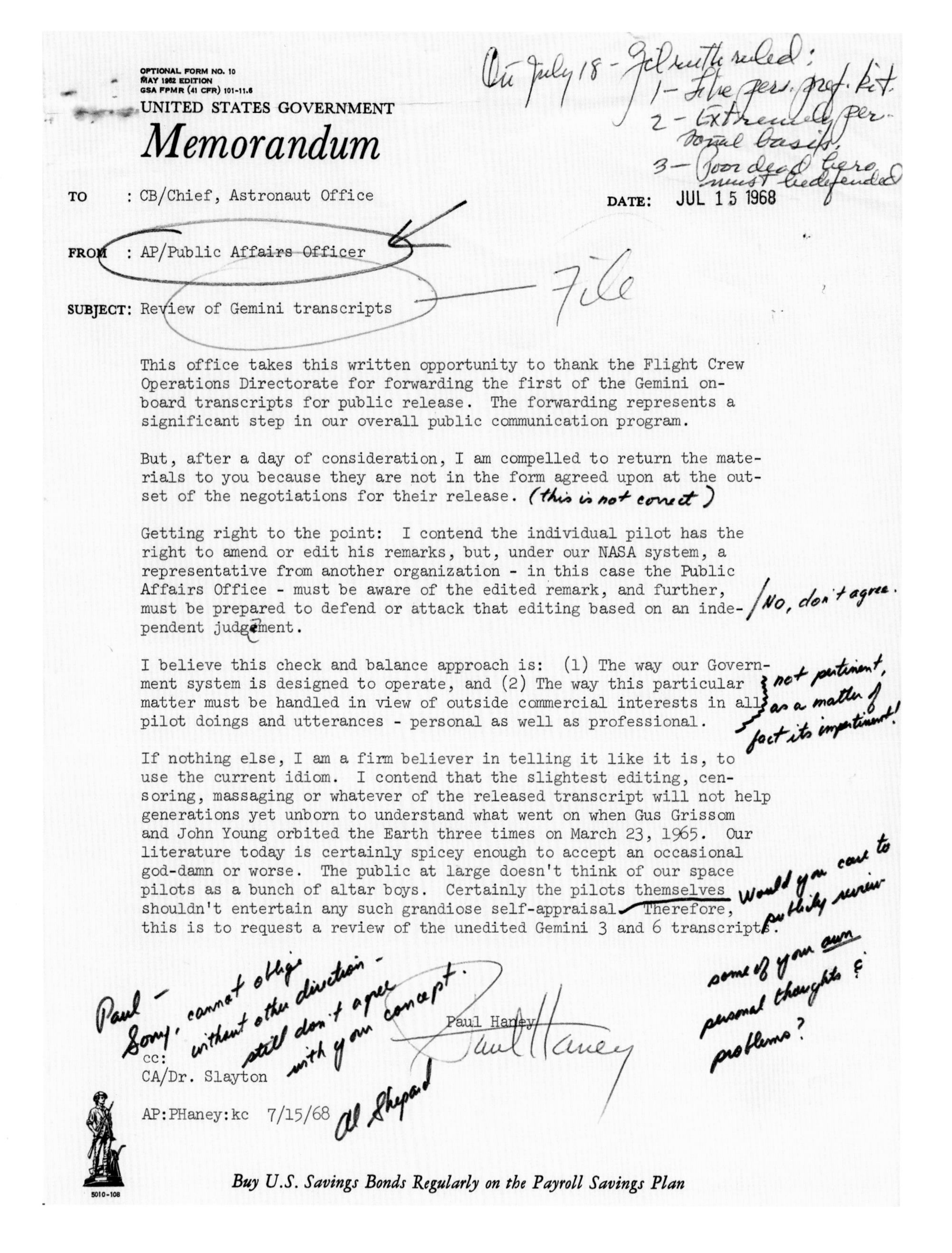

OPTIONAL FORM NO. 10
MAY 1962 EDITION
GSA FPMR (41 CFR) 101-11.6
UNITED STATES GOVERNMENT

Memorandum

On July 18 - Gilruth ruled:
1 - Like pers. pref. kit.
2 - Extremely personal basis.
3 - Poor dead hero must be defended

TO : CB/Chief, Astronaut Office   DATE: JUL 15 1968

FROM : AP/Public Affairs Officer

SUBJECT: Review of Gemini transcripts — File

This office takes this written opportunity to thank the Flight Crew Operations Directorate for forwarding the first of the Gemini on-board transcripts for public release. The forwarding represents a significant step in our overall public communication program.

But, after a day of consideration, I am compelled to return the materials to you because they are not in the form agreed upon at the outset of the negotiations for their release. (this is not correct)

Getting right to the point: I contend the individual pilot has the right to amend or edit his remarks, but, under our NASA system, a representative from another organization - in this case the Public Affairs Office - must be aware of the edited remark, and further, must be prepared to defend or attack that editing based on an independent judgement. / No, don't agree.

I believe this check and balance approach is: (1) The way our Government system is designed to operate, and (2) The way this particular matter must be handled in view of outside commercial interests in all pilot doings and utterances - personal as well as professional. } not pertinent, as a matter of fact its impertinent!

If nothing else, I am a firm believer in telling it like it is, to use the current idiom. I contend that the slightest editing, censoring, massaging or whatever of the released transcript will not help generations yet unborn to understand what went on when Gus Grissom and John Young orbited the Earth three times on March 23, 1965. Our literature today is certainly spicey enough to accept an occasional god-damn or worse. The public at large doesn't think of our space pilots as a bunch of altar boys. Certainly the pilots themselves shouldn't entertain any such grandiose self-appraisal. Therefore, this is to request a review of the unedited Gemini 3 and 6 transcripts.

Would you care to publicly review some of your own personal thoughts & problems?

Paul Haney

Paul - Sorry, cannot oblige without other direction - still don't agree with your concept. Al Shepard

cc:
CA/Dr. Slayton

AP:PHaney:kc 7/15/68

5010-108

Buy U.S. Savings Bonds Regularly on the Payroll Savings Plan

Lyons—which ran in many newspapers under headlines such as "Astronauts Stand to Share $1 Million in Story Rights," "America's Spacemen A Well-Heeled Lot," and "Astronauts' Exploits Mean Cash in the Bank"—Haney was quoted saying, "one of the fundamental reasons for the contract with *Life* was the $100,000 life insurance contract each man received." The other reason, said Haney, was to provide the astronauts "leverage against NASA" so they could avoid cooperating with Public Affairs on interview requests.[33]

The soap-opera news cycle that followed Haney's resignation soon faded from the headlines as the launch of Apollo 11 drew near, but its after-effects cast a pall over what was one of the more notable careers in NASA Public Affairs. "Haney had been down there five years. He was colorful and he was a household name," said Duff. "His problem was that he'd gotten out of sync with his bosses in Houston, and, in doing so, with Washington, too."

Indeed, just a month after Haney's resignation and run-in with Scheer, it was reported in a syndicated Washington column titled "Capital Chatter" that "NASA Public Relations Chief Julian Scheer . . . discovered that hell hath no fury like congressmen scorned. In a fit of pique, the House Space Committee slashed $1.5 million from the $3.6 million Public Affairs budget submitted by Scheer" because "the congressmen had their noses out of joint because they and some of their important constituents hadn't been invited to witness some of the recent Apollo launchings." And while the article mentioned that the reassignment and subsequent resignation of Haney had nothing to do with the slashing of the budget, the article implies otherwise in a mention that the funding would "likely be restored in House-Senate conference, but the House committeemen figure the point won't be lost on Scheer."[34]

It wasn't lost on Duff, either.

"Haney's problem was that he forgot that, no matter how big and how wonderful and how competent the people were in Houston, they were still part of the larger organization. Houston was not the space program, NASA was the space program, and NASA was still run by Washington," recalled Duff. "The second thing I perceived about Haney's problem was that he got too visible. People who'd never heard of Bob Gilruth knew Paul Haney. Some people called him the 13th astronaut or something like that. That's death."[35]

Getting in Line with Washington

Upon arriving in Houston, Duff created a rotating duty roster for voicing mission commentary out of MSC, appointing veteran Houston PAO staffers Jack Riley, Doug Ward, and Robert T. White to the assignment, a plan that Riley publicly commented would hopefully "shade himself from the public spotlight associated with the job in the past."[36] There would no longer be a cult personality serving as the "voice of NASA." It was an important turning point for Public Affairs, a time in which the field began to operate more and more in line and in sync with Washington.

"I made a very deliberate effort to create the appearance of seeking anonymity, seeking a low profile, and pushing the staff affairs people who were actually working with the programs into a position of visibility," Duff later recalled, explaining how he adapted to the MSC environment after taking over from Haney. "I never went on the air. I took the position that I was a member of the senior staff of the Manned Spacecraft Center, and I didn't have time to be a radio announcer. That was another very deliberate effort . . . I didn't run very many press conferences. I was trying to make the impression that I was the manager of this function," rather than one of its star personalities, and "I think my tactics worked."[37]

The Formative Years

About the time Haney was replaced by Brian Duff, the Public Affairs Office in Houston had grown to a team of some sixty people, including thirty-five regular Public Affairs staffers and thirty-five "special assignment" contract employees, and it operated on a budget of roughly $500,000 a year.[38] This was a surprisingly small budget and staff roster, given that roughly 90% of all NASA PR activity at the time was focused on Houston. It was the epicenter of activity and interest for the entire length of the Apollo program—only usurped during the those few days when Cape Kennedy and the Kennedy Space Center held sway for the actual launches. By the late 1960s, the Manned Spacecraft Center had become a national attraction. It drew more than 800,000 visitors in 1968. In comparison, 500,000 people visited the Grand Canyon the same year.[39]

The MSC in Houston was a field office, and NASA had eleven of them around the country in addition to NASA headquarters in Washington, D.C. During the Apollo program, there was

The May 18, 1962, Life *magazine cover story of Scott Carpenter and his first wife, Rene, is an example of the personal stories to which the magazine acquired exclusive right throughout most of the Mercury, Gemini, and Apollo programs.*

33. Richard Lyons, The New York Times Service, "Astronauts Stand to Share $1 Million in Story Rights," June 22, 1969.
34. "Capital Chatter," *The Press-Telegram*, Long Beach, California, May 22, 1969.
35. Brian Duff, Smithsonian Air and Space Museum Oral History Interview, 1989.
36. "New 'Voice of Apollo' Wants Out of Spotlight," United Press International, May 17, 1969.
37. Brian Duff, interview, 1989, op. cit.
38. Will McNutt, "Voice of Apollo Goes Off The Air," *Sun Science Wire Service*, May 18, 1969.
39. Ibid.

Are You a Turtle?

The flight of Apollo 7, in October 1968, brought the program back from the long period of recalibration following the tragic fire of Apollo 1 , twenty-one months earlier. The mission had the attention of the world, as it was the first from which live television broadcasts would be made. Millions of viewers tuned in to watch it. While the broadcasts were black and white and very grainy, audiences were treated to a tour of the spacecraft and a little bit of earth-to-ground hijinks by the astronauts that garnered its own share of publicity. One stunt, in particular, made its way into newspaper articles and television broadcasts around the world and contributed, in no small part, to a growing sense by Julian Scheer, that Paul Haney, head of Public Affairs at the Manned Spacecraft Center in Houston, was becoming more an independent operator than a public servant. It became known as the "Turtle" incident.

The Ancient and Honorable Order of Turtles, known also as the International Association of Turtles, was a game that is said to have originated with pilots during World War II. What began as an inside joke grew into a "club" with chapters and members around the world numbering in the hundreds of thousands, perhaps millions. It had one activity: once inducted, a member was required to respond to the question, "Are you a Turtle?" with, "You bet your sweet ass I am!" If, for whatever reason, the member failed to give the reply, then he (it was invariably a he) had to buy a round of drinks for all Turtles present. The goal was to test the resolve and courage of the member in a situation in which the official reply might cause embarrassment.

During the Mercury days, astronauts Alan Shepard and Wally Schirra formed an "intergalactic" chapter of the Turtles. Only three-and-half minutes into Schirra's maiden trip into space on *Sigma 7*, capsule communicator Deke Slayton asked his fellow astronaut, "Are you a turtle today?" Schirra quickly responded that he was switching to the on-board voice recorder to leave his answer rather than broadcast it over the live audio network. (NASA mission transcripts of the 1962 on-board recorder do not note Schirra's words, only "correct answer recorded.") So when preparing for his Apollo 7 mission six years later, Schirra decided to get his revenge on Slayton during the October 15 live television broadcast. Working with a friend in the broadcast industry, Schirra had a series of cue cards prepared on fireproof paper to use as part of his air-to-ground conversation during the broadcasts. Among the cards were two Turtle call-outs: one to Deke Slayton, now the Director of Flight Crew Operations, and the other to Paul Haney. The cards had a picture of a turtle and "the question," asked directly of both Slayton and Haney, "Are you a Turtle?" Slayton recorded his answer on a tape recorder, mirroring what Schirra had done. Haney did not respond at all. Gene Cernan, who served as capcom for the mission, informed Schirra, "Wally, this is Gene. Deke just called in and we've got your answer, and we've got it recorded for your return." Later, when talking with capcom Jack Swigert, Schirra asked if Haney had provided an answer. "No, Haney isn't talking, Wally," Swigert said, "Somebody tells me he isn't talking, but just buying."

The incident was covered by news outlets all over the world, and Haney's public non-response spurred only additional unwanted press attention and notoriety. Haney was summoned into the office of Manned Spacecraft Center Director Robert Gilruth to explain what was going on, and was soon inundated with hundreds of letters from Turtles demanding that he buy them a drink.

And with instant celebrity came some surreal moments. While the Apollo 7 mission was still in progress, Haney was seated at his Houston console when his private phone rang. "Hi. This is Bob Hope," said the voice on the other end of the line. Thinking this was yet another prank, Haney started arguing with his caller. Aware that the real Bob Hope spent part of his youth in Ohio, Akron-born Haney needled him with

This television screen shot of one of the "Are You A Turtle?" cue cards flown on Apollo 7, was printed here on the front page of the New York Times, *as it was in newspapers around the world.*

MOON SHOT HOPES RAISED BY APOLLO

Continued From Page 1, Col. 7

the Apollo 7 flight director, and Paul Haney, the space center's public affairs director, did not deny that such plans had been made.

They tried to duck the whole subject, saying, "No one tells us anything."

"Coming to you live from outer space, the one and only original Apollo Everything Road Show," said Major Eisele, opening the television sequence which was seen live on national television. "Starring those great acrobats of outer space, Wally Schirra and Walt Cunningham."

At that moment, Captain Schirra and Mr. Cunningham were out of their couches and floating in front of the camera.

Captain Schirra was holding up another of his signs. This one read: "Deke Slayton, are you a turtle?"

Donald K. Slayton is director of flight crew operations. The question is supposed to elicit in response the password of an informal society, mainly composed of air space people, called the Turtles.

The society's one rule is that any time one is asked "Are you a turtle?" the member must reply "You bet your sweet—— I am." If the person fails to answer in full, he must buy drinks for everyone within earshot.

Mr. Slayton insists he recorded the answer on tape for the astronauts when they return.

Major Eisele then held up a sign: "Paul Haney, are you a turtle?"

Mr. Haney, whose voice was going out from the control room on the television networks, did not answer.

"You mean he's speechless?" asked Major Eisele, apparently ready to collect a drink.

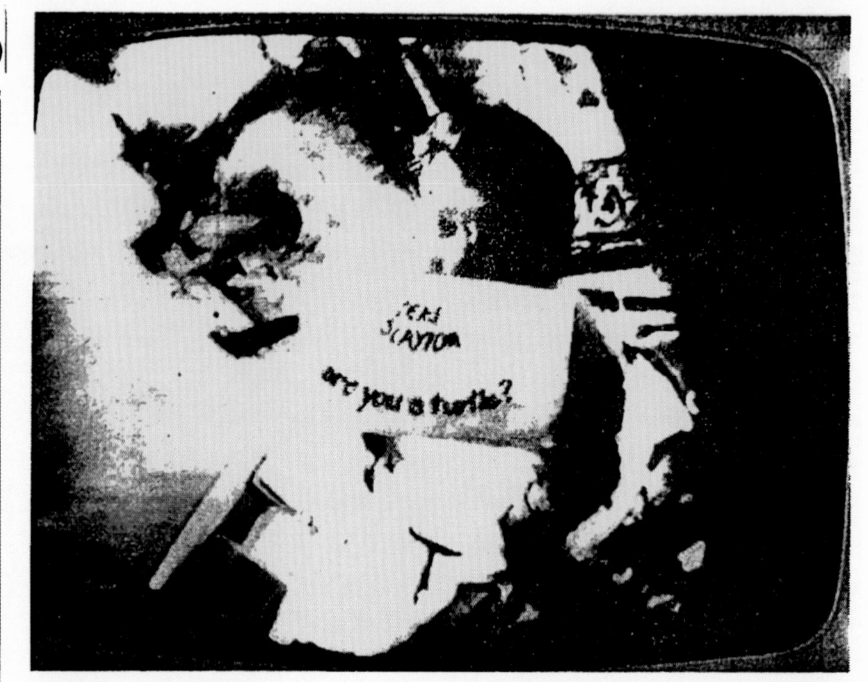

Associated Press

'THE APOLLO EVERYTHING ROAD SHOW,' as Maj. Donn F. Eisele of the Air Force called it yesterday. Walter Cunningham was on camera as the spacecraft televised live for ten minutes yesterday from high over the Gulf Coast.

specific questions about the Buckeye State. Detailed answers about Cleveland's Euclid Avenue eventually convinced Haney that this was no hoax. "Hi, Bob," Haney said as he recovered his professional composure, "What can I do for you?"

Hope was preparing an NBC variety special that was to include a salute to the 10th anniversary of the astronaut program and feature both Hollywood stars and real-life astronauts. So, on the evening of Wednesday November 6, 1968—the day after America elected Richard Nixon the next President of the United States—viewers of NBC saw Bob Hope visiting the Manned Spacecraft Center in Houston, where he introduced the newly returned Apollo 7 crew and Turtle celebrity Paul Haney to actress Barbara Eden, then famous for playing the romantic partner of fictional astronaut Tony Nelson (Larry Hagman) on the popular sitcom *I Dream of Jeannie*. (Hope's program also featured music from Ray Charles and a satire of the then current hit movie *The Odd Couple*, in which Hope and actor David Jansen played a mismatched pair of astronauts who get on each others nerves after 97 days in space.) Haney later remarked, "It was a good show . . . [however] Hope didn't make me any new friends at NASA in Washington."

Turtle organizations worldwide reported an avalanche of applications for new members, with one organization in the United States claiming more than 700,000 members by 1971. Later, Schirra presented the flown cue card to Haney as a memento. It was on display for many years at the Smithsonian's National Air and Space Museum.

Seven months after the Turtle affair, Haney resigned his post as MSC head of Public Affairs.

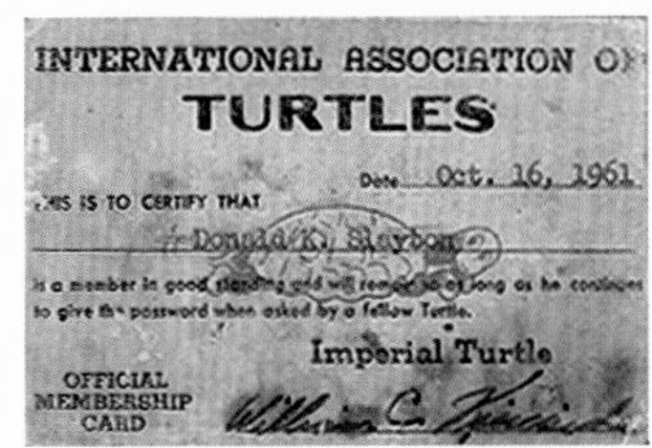

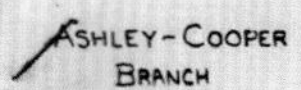

Ashley-Cooper Branch

Charleston South Carolina

International Association of

Turtles

October 16, 1968

Capt. Walter M. Schirra, Jr.
Manned Spacecraft Center
Houston, Texas

Dear Fellow Turtle:

First, congratulations to you, Walter Cunningham and Donn Eisele on your outstanding mission onboard Apollo 7.

Now, on behalf of the Ashley-Cooper Turtles and all Turtles throughout the land, I would like to thank you for making us interplanetary in addition to being international. The VOICE OF THE TURTLE HAS BEEN HEARD from outer space on October 15, 1968. Because of Apollo 7 we are now known all over this Earth and not just on the Muddy Bank of Turtle Creek.

Shame on Paul Haney for not being a Turtle in good standing and responding the way he should have. Also, these people who said, "if properly answered, would have elicited a racy reply not fit for the airways" - this just goes to show that they are not of clean mind as all of us good Turtles are.

I would like to take this opportunity to invite you and the other members of Apollo 7 to the Ashley-Cooper Turtles' Christmas Party which will be held on Friday, December 13th. Enclosed are membership cards for the three of you. Most of the Ashley-Cooper Turtles are active or retired Naval personnel.

Keep up the good work in spreading Turtledom and here's hoping that Apollo 8 will have the first Turtle on the Moon this December. This is undoubtedly the Year of the Turtle.

In closing fellow Turtle, I would like to say God bless you and all the astronauts in helping keep America strong and safe.

A Turtle always,

Ron Fledderman
Ron Fledderman
Imperial TURTLE
2430 Falcon Road
Charleston Heights, S.C. 29405

P.S. Are you a TURTLE?

cc:
Mr Paul Haney

LEFT: *Paul Haney proudly displays the flown "Are you a Turtle?" cue card in his home office.* **TOP:** *Paul Haney makes an appearance on television with Bob Hope, Barbara Eden, and the crew of Apollo 7, further raising his own profile in the press.* **ABOVE, CENTER:** *Astronaut Deke Slayton's Turtle club membership card, and a close-up of the Paul Haney cue card flown on the mission.* **RIGHT:** *One of the many letters received by Wally Schirra and NASA from Turtle organizations around the world after the "Are you a Turtle?" broadcast gag.*

TWA was awarded a contract to provide base support to the Kennedy Space Center in 1964. The contract incuded general maintenance, utilities, supply operations, fire prevention and protection, security, and medical services. In 1966, the contract was extended to inlcude on-site visitor tours. During the life of the contract, more than 2,000 TWA employees worked for NASA at KSC, and millions of visitors took a TWA tour during the Apollo era.

a natural tension between the field offices and headquarters, in many cases driven by the complexity of such a large agency and the nature of their respective functions within the organization. "Houston was the consummate field center," explained Duff. "It had the real job of flying human beings in space, overseeing building the spacecraft and the systems, controlling them, and training the astronauts. Washington had the dirty job of getting the money and gaining the public support, interfacing with the rest of the huge federal government . . . doing the infighting with the other agencies that were always after our budget, or our people, or our prestige—our access to the president or the key chairmen."[40]

It is a dynamic that had been with NASA literally since the time of its formation, in 1958, when it evolved out of the nearly fifty-year old National Advisory Committee on Aeronautics (NACA), and a main catalyst behind the slow and rocky evolution of the policies and influence of Public Affairs.

"NASA, I like to point out to people, was sort of put together at a federal yard sale on a Saturday," Haney once explained. At its founding, NASA assumed more than 7,500 NACA employees and $300 million worth of research field facilities, including Langley, Lewis, and Ames,[41] as well as many functions that, up until that time, were considered the exclusive domain of the Department of Defense. NASA was the new civilian kid on the block, put in charge of what for decades had been the very private and seriously secretive responsibility of the military. It was neither a smooth, nor a very easy, transition and the interagency tensions had a strong impact on Public Affairs policies throughout the early years. "This was supposedly a marriage arranged in downtown Washington," said Haney, "but boy, those are tough marriages."[42]

Walter T. Bonney and Open Public Relations

Tough marriage, indeed. Even at headquarters, the public affairs function—originally called "public information"—had its mission and roots embedded in the National Aeronautic and Space Act of 1958, which required the agency to "provide for the widest practicable and appropriate dissemination of information concerning its activities and results."[43] The act also included a provision much like the Freedom of Information Act, which mandates the disclosure of information obtained or developed in conjunction with its activities.

The American space program conspicuously sought to set itself apart from Russia's covert aerospace activities, by promoting an "open program" approach to information. In theory, it was a simple mission. In practice, it was a heavy task to juggle the conflicting and often contradictory public affairs agendas of Washington, the military, NASA's headquarters and field offices, the astronauts, and the press.

From the beginning of the program, Walter T. Bonney took his charge seriously, establishing some of the agency's first major and important policies. One of his early priorities was to ensure that NASA, and not the competing areas of the Army and Air Force space programs, served as the primary source of information to the public. "From the outset, NASA must maintain a positive information program designed to provide the people of the United States with maximum information about the agency's accomplishment," Bonney wrote to NASA administrator T. Keith Glennan. "We should speak of our work modestly but with enough vigor to be heard. If we don't, others soon will clamor to take our space assignment from us."[44]

Headquartered in the old Dolley Madison Building, in Washington, NASA held one of its first high-profile public events when Bonney introduced the Mercury 7 astronauts in an auditorium converted from a former horse stable.[45] In those early days, every move down at Cape Canaveral—every missile launch, every dropped bolt—was observed by a curious press corps, and it created a tremendous amount of work

40. Brian Duff, interview, 1989, op. cit.
41. Mark Erickson, *Into the Unknown Together: The DOD, NASA, and Early Spaceflight.* Maxwell AFB, Ala.: Air University Press, 2005.
42. Haney, Oral History Interview.
43. National Aeronautic and Space Act of 1958.
44. Ginger Rudeseal-Carter, op. cit.
45. Ibid.

(a) If a (GA) - ASTRONAUT is in the Military Forces and is dishonorably discharged by his Department, or if a (GA) - ASTRONAUT whether a civilian or a member of the Military Forces is discharged for cause, other than mental or physical illness, he shall immediately forfeit all amounts which may thereafter be distributed.

6. In the event of the death of a (GA) - ASTRONAUT at any time while he is entitled to participate in the proceeds as provided by this Agreement, any sums thereafter received, until the completion of the Projects, which would otherwise be payable to him shall be paid to the personal representative of his estate.

7. Harry A. Batten agrees to serve the (GA) - ASTRONAUTS without any compensation, and will personally defray all expenses incurred by him in connection with his representation. The agency of Harry A. Batten may be terminated at any time (a) by agreement between Harry A. Batten and a majority of the (GA) - ASTRONAUTS, or (b) in the absence of such agreement, by the concurrent action of a majority of the (GA) - ASTRONAUTS. In the event of such termination, Harry A. Batten agrees to re-assign to the (GA) - ASTRONAUTS all of their rights, subject, however, to such commitments as may have been made on their behalf by the said Harry A. Batten prior to the termination of his agency.

IN WITNESS WHEREOF, the parties hereto have hereunto set their hands and seals as of the day and year first above written.

Edwin E. Aldrin, Jr. (SEAL)	Eugene A. Cernan (SEAL)
William A. Anders (SEAL)	Roger B. Chaffee (SEAL)
Charles A. Bassett, II (SEAL)	Michael Collins (SEAL)
Alan LaV. Bean (SEAL)	Ronnie W. Cunningham (SEAL)

The undersigned, being the wives of the (GA) - ASTRONAUTS, parties to the foregoing Agreement, hereby consent to and join in the said Agreement as their interests may appear.

Joan Ann Aldrin (SEAL)
Valerie Elizabeth Anders (SEAL)
Jean Martin Bassett (SEAL)
Sue Ragsdale Bean (SEAL)
Barbara Jean Cernan (SEAL)
Martha Horn Chaffee (SEAL)
Patricia Mary Collins (SEAL)
Loella Cunningham (SEAL)
Harriet Elaine Eisele (SEAL)
Faith Clark Freeman (SEAL)
Barbara Field Gordon (SEAL)
Clare Whitfield Schweickart (SEAL)
Anne Lurton Scott (SEAL)

UPPER LEFT: *Shorty Powers with Leo DeOrsey, the Washington, D.C. attorney and president of the Washington Redskins football team, who was selected by Walter Bonney to negotiate and manage the astronauts' personal public relations affairs.*

CENTER AND RIGHT: *The Public Relations Contract. Walter Bonney, NASA's first head of public affairs, is credited with the idea of granting exclusive rights to the astronauts' personal stories to a single outlet—*Life *magazine, later* Life*/World Book Science Services—during the early Mercury program. Bonney suggested it as a way to control the media's intrusion on the astronauts' personal and family lives. The astronauts would need an agent to collectively negotiate on their behalf and serve as advisor for any and all such arrangements. Bonney selected Leo DeOrsey, who represented the astronauts pro bono, and served the Mercury 7 and Group 2 astronauts. When DeOrsey died in 1965, the Mercury 7 astronauts were pall bearers at his funeral. In March 1964, the Group 3 astronauts, including Buzz Aldrin and Michael Collins of Apollo 11, as well as fellow moonwalkers Dave Scott, Alan Bean, and Gene Cernan, selected Henry Batten to represent them. Batten, a Philadelphia native and the head of the N. W. Ayer & Son advertising agency, remained the astronauts' agent and advisor on business and personal affairs, working pro-bono like DeOrsey, for the next two years until his death at age 69, in July 1966. In October of that year, the astronauts selected Louis Nizer, a well-known New York City attorney, as their advisor and agent. Shown here are the signature pages of the agreement with Batten. Note that the astronauts' wives were also parties to the agreement.*

LEFT: *National Advisory Committee for Aeronautics (NACA) sign being taken down as the agency's name is changed to NASA.*

RIGHT: *T. Keith Glennan, the new NASA Administrator, and Hugh L. Dryden (left), Deputy Administrator, are sworn in August 19, 1958, as President Eisenhower looks on.*

for the small office. "With a modest staff, Bonney met the full thrust of the enormous pressure of the news media during the hectic early years of the space race," wrote Eugene Emme, NASA historian. "NASA's projects, inherited from . . . the military services, plus Project Mercury, found Bonney at his NASA desk from dawn to dusk every day."[46] By 1960, Bonney was overseeing a staff of twenty-eight Public Information officers, logging hundreds of overtime hours, issuing almost 800 press releases a year, and distributing thousands of pictures to the world media,[47] and it still wasn't enough to keep up with the public demand for information.

Bonney, who had been a reporter for a small newspaper in central Illinois, sought to have his staff "work as reporters" and to prepare materials "in such form and content as will be useful to the general press, trade press, radio, television, magazines, and writers of non-technical books about space."[48] He pushed for school education programs, helping to launch the Space Mobile program, and for more public speaking by NASA employees. Bonney also felt strongly that his office should have a hand in the annual and semi-annual NASA reports to Congress, as he viewed these documents as "sell presentations" that needed direct input from the Office of Public Information. He also pushed, and received, expanded influence beyond the home office, establishing Public Information field offices and personnel at the major centers. One important posting was Jack King, who became Cape Canaveral's first public information officer. He was responsible for coordinating NASA's interaction with the press, which had taken interest in the growing number of missile launches—many of which, in the early days, ended in dramatic, fiery failures over the ocean.[49]

Try as he might, the interagency tensions and the growing pressures of building procedural and infrastructure functions of NASA Public Information would ultimately prove to be too much for Bonney—and for the several other Public Affairs heads who followed him. He would leave in the fall of 1960 to join an aerospace company in California, after more than fifteen years with NACA and two years with NASA.[50] Between 1960 and 1963, NASA Headquarters had five public information directors: Bonney, Shelby Thompson, O.B. Lloyd (who served in the post twice), Hiden T. Cox, and George Simpson.

The head of NASA Public Affairs in Washington had become a carousel of leadership. However, most of the staff remained consistent: Jack King at the Cape, and Shorty Powers at the MSC in Houston, as well as Al Alibrando, Paul Haney, Harry Kolcum, Dick Mittauer, Herb Rosen, Joe Stein and other influential early information officers. They provided the program with stability in execution and a constant flow of information to the press and public throughout the early years.

But the rapid, almost constant change in leadership meant that NASA Public Affairs was diminished in its ability to influence and persuade on a national level. "Public affairs is a staff function, in which you recommend," said Duff, explaining the often delicate, diplomatic role Public Affairs officers had within the NASA agency. "You do not direct, you recommend, and you counsel, and so therefore you have to persuade." Without stable leadership, the influence of Public Affairs on shaping

46. Eugene Emme, Biography of Walter T. Bonney, as quoted in Rudeseal-Carter, *Public Relations Enters The Space Age,* Rudeseal-Carter, op. cit.
47. Ginger Rudeseal-Carter, op. cit.
48. Walter T. Bonney, "Memorandum for the Administrator, NASA," Washington, D.C., September 9, 1958.
49. Ginger Rudeseal-Carter, op. cit.
50. Harlen Makemson, op. cit., p. 57.

NASA's story and its public image—particularly in Washington, but also nationally—would remain in a weakened state until the appointment of Julian Scheer in 1963. Scheer would hold the job for eight years, through most of the Apollo lunar landing missions.

Julian Scheer: Live and Real Time

On February 3, 1963, NASA Administrator James Webb appointed Scheer to head up NASA Public Affairs in Washington. He was tough reporter for the *Charlotte News,* who had previously covered politics, the early civil rights movement, and the space program. His appointment would represent a dramatic change in influence. Prior to his appointment, Webb had hired Scheer as a consultant, to draft a proposal on how to improve the Public Affairs Office and project a better image of NASA.[51] After a few months, Webb convinced Scheer to take on the administrative responsibilities full time.

"We are going to get information out, and we are going to tell the truth," Scheer told a journalism trade publication, early in his tenure at NASA. "It's ridiculous to have an information program predicated on anything else. I complained like everybody else when I was reporting and I couldn't get information that officials were withholding. I didn't like it, and I am not going to be a party to that kind of stuff now."[52]

Brian Duff asserted that, when Scheer took over the Public Affairs position, he put a stamp on it that lasted for decades. Scheer is most often credited with being the NASA Public Affairs chief who most embraced the open program concept, and was "probably the most respected public affairs officer in Government because his office and NASA in general" operated so openly.[53] Even though that concept was put through strenuous tests at times—most notably during the poor handling of information about the Apollo 1 fire—it served as a model of transparency during the dramatic events of the Apollo 13 mission, as NASA "quickly and frankly told the world of the problem, and kept media informed for the next four days as engineers and flight controllers devised a plan to return the astronauts safely to Earth."[54]

With the direct support of Webb, Scheer exerted more control and influence over the field offices. "Julian came in and put his fist down," Jack King recalled. "The atmosphere changed. He gave us the clout out of headquarters," and it facilitated a much smoother working function both in the field and between the field and headquarters. "Julian ran a tight ship and, just from an information point of view, he was really the key guy in Public Affairs throughout the whole program, all the way up to Apollo. He and Webb got along great, and Webb was a tremendous administrator."[55]

Brian Duff described Sheer as a very strong personality. "He didn't pull his punches. He had no favorites."[56] In Scheer's early analysis of the program, he noted staffers making efforts to pry "information out of reluctant hands," a legacy of the space program's birth in the military. There was also "little or no unity or coordination of action among the various elements of Public Affairs. The field centers were highly autonomous and received little or no guidance or direction from headquarters, except in a crisis. The substantive program was largely reactive and consequently wasteful. Headquarters operated very much like another field center—instead of making policy, providing direction and guidance, and coordination."[57] Working with Webb, Scheer set about to correct this imbalance, and

One of the more innovative and enduring educational programs funded by NASA was the Space Mobile, a museum-on-wheels staffed by drivers who crisscrossed the nation in their "Space Mobile" station wagons and vans, bringing models and displays to schools and auditoriums to give lectures about NASA and space exploration. It not only sought to influence the youngest minds of the country about the exciting adventures being conducted in space by NASA, but also to inspire students to study science and engineering. From its earliest days, the program, also known as the Aerospace Education Services Program, was outsourced to third-party vendors, such as Oklahoma State University and Penn State University.

51. Ibid.
52. Caryl Rivers, "NASA: News Chief Respects Reporters," in *Editor and Publisher*, April 4, 1964, p. 44.
53. Journalist quoted in Harlen Makemson, op. cit., p. 202.
54. Ibid.
55. Jack King, interview with the authors, November 9, 2011.
56. Brian Duff. Smithsonian National Air & Space Museum Oral History Interview. May 1, 1989.
57. Harlen Makemson, op. cit., p. 101.

TOP: *Julian Scheer*
ABOVE: *Jack King*

in doing so, removed Powers and then Haney. "Those who bucked the system, either inside or outside the agency, quickly discovered that Scheer had the complete backing of his boss," noted historian Harlen Makemson.

"You can't give Scheer enough credit for his belief in what we called the 'open program' and his ability to convince NASA management that this was in the best interests of the agency," said Duff.[58] Yet despite all the laurels heaped upon Scheer, his underlying philosophy echoed that of Walter Bonney: it was not the business of NASA Public Affairs to "spin" information, but to inform and educate the public in an accurate and timely manner. Overt attempts at marketing the Apollo program were met by Public Affairs leadership with disdain. "We are not doing what is known in the public relations business as flackery or publicity or public relations or propaganda," Scheer once said, outlining his philosophy. "We are simply not in this kind of business. We are not buying refreshments, we are not supplying free trips, we are not slapping anyone on his back, we are not spending our time at the press club bar. We feel that we have a service to offer and we offer our service as best we can and we have to stand on that performance. Therefore, what we are in public information is a news operation. We don't put out publicity releases. We put out news releases. When we have news, we disseminate it."[59]

Scheer's approach is one that worked well in the time of Apollo, when the world embraced the drama and adventure of the great competition between the two superpowers. It would present challenges in the future—challenges that some contend NASA failed to properly address from a marketing and PR perspective. But during the summer of 1969, it functioned to perfection. Just in time for Apollo 11, Scheer compiled a comprehensive public affairs plan that would insure everyone knew his place and duty and that his once-wayward department would function like a clock.

What Scheer would not or could not do, would and could be done by the contractors, the thousands of manufacturing and service companies, including some of the nation's largest firms, who would work more closely with NASA than did any government-private sector partnership before or since. The approach made sense, as only the contractors could provide the kind of high-level information that was increasingly demanded by the public and the press.

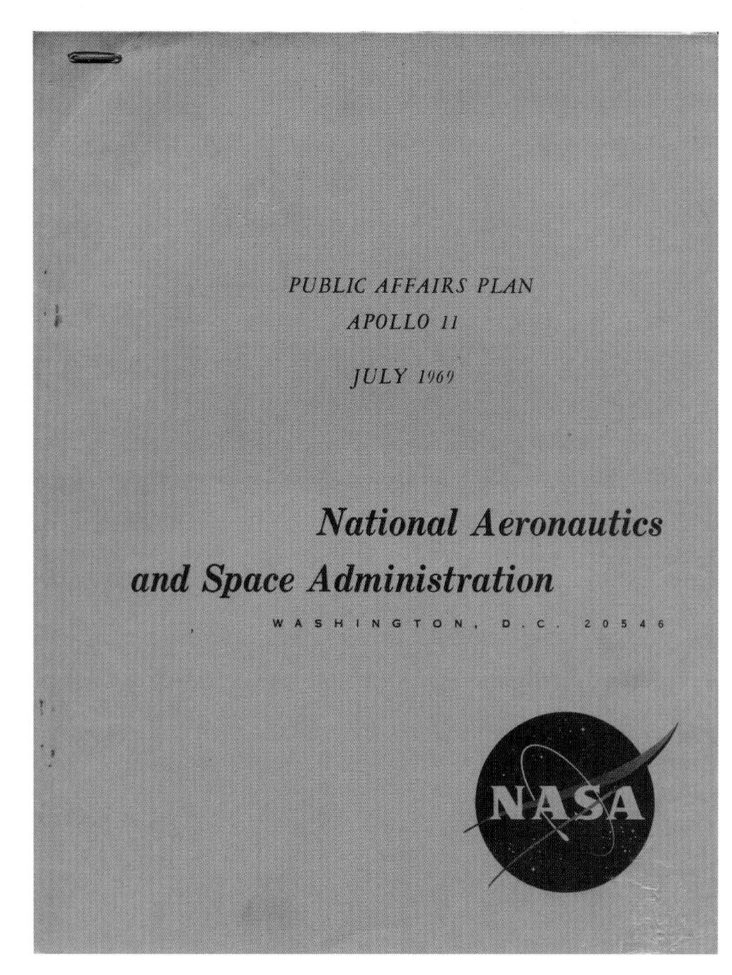

PUBLIC AFFAIRS PLAN
APOLLO 11
JULY 1969

National Aeronautics
and Space Administration
WASHINGTON, D.C. 20546

"The world was beginning to hear astronauts in real time," Scheer said. "It was an evolutionary process of openness which I think culminated when astronauts walked on the Moon. Here the whole world could see one of the great events of all time. Live and real time, unedited. And I think all of us felt tremendous pride in being able to do that. And it was a product of a lot of hard work on the part of Public Affairs and a lot of skill and the organization and from the support of the technical people, who began to realize that this was what it was all about."[60] ⊙

58. Brian Duff, oral history, 1989, op. cit.
59. "Public Affairs Program Review Document," April 19, 1966, NSA History Office, Washington, DC. Cited by Makemson, op. cit., p. 103.
60. Ginger Rudeseal-Carter, op. cit.

NATIONAL AERONAUTICS AND SPACE ADMINISTRATION
WASHINGTON, D.C. 20546

OFFICE OF THE ADMINISTRATOR

JUL 1969

MEMORANDUM to The Staff, Office of Public Affairs

Attached is the Apollo 11 Public Affairs Plan.

It is not intended as a comprehensive document reflecting all of the four divisions of this office (Public Information, Special Events, Educational Programs, Media Development) and their work specifically related to Apollo 11 or the continuing work of the divisions in the total NASA Public Affairs program. Special materials developed for Apollo 11 are mentioned only briefly. Post-flight plans are incomplete. It should also be noted that every public affairs office in all 14 locations in the United States contributed in part to the development of this Plan and to the program.

This document does reflect the major operational procedures and is subject to change, amendment and clarification.

This document, in spelling out policy, procedures and practices, does reflect the open program conducted by this Agency. This is not subject to change under any circumstances.

Your continued support is required; your dedication is appreciated.

Julian Scheer
Assistant Administrator
for Public Affairs

Attachment

2

Apollo 11 will carry an Apollo Lunar Radioisotopic Heater (ALRH) to provide heat for the passive seismometer in the EASEP to permit its operation through at least one lunar night. The heater contains about two ounces of plutonium 238 as the heat source.

ORGANIZATION AND RESPONSIBILITIES

The NASA Assistant Administrator for Public Affairs (Julian Scheer) is responsible for all NASA public affairs activities. His assistant for NASA-wide public information functions is the Director, Public Information Division, NASA Headquarters (O.B. Lloyd, Jr.).

For Apollo missions, responsibility for direction of public information activities has been assigned to the Public Affairs Officer, Office of Manned Space Flight (PAO/OMSF), NASA Headquarters (Alfred P. Alibrando). In this capacity he will coordinate his activities with and be responsive to the Apollo Mission Director when his duties impact mission operations.

In implementing this plan, the PAO/OMSF is designated Apollo Information Director. He will be responsible for the organization, coordination and accomplishment of all public information activities pertaining to the Apollo missions. He has delegated responsibility as follows:

a. Assistant Public Affairs Officer, OMSF (William J. O'Donnell) is designated Deputy Apollo Information Director (Information). He will assist the Apollo Information Director in the overall direction of the mission Public Affairs program.

b. NASA Public Affairs Officer (Richard T. Mittauer) is designated Deputy Apollo Information Director (Operations). He will assist the Apollo Information Director in developing and carrying out the Apollo 11 Public Affairs Plan.

c. Public Affairs Officer, MSC (Brian Duff), designated the Deputy Apollo Information Director (MSC), will be responsible for the following:

(1) Establishing and managing the MSC Apollo News Center.

(2) Scheduling briefings and press conferences at MSC.

(3) Providing the Apollo mission commentary from the Mission Control Center as directed by the Apollo Information Director.

(4) Providing a transcript of the Apollo mission commentary to the MSC and KSC News Centers.

(5) Providing flight crew status to KSC and MSC News Centers prior to launch.

3

(6) Processing and releasing onboard spacecraft photography.

(7) Post-mission information operations at the Lunar Receiving Laboratory (LRL).

d. Public Affairs Officer, KSC (Gordon L. Harris), designated Deputy Apollo Information Director (KSC), will be responsible for the following:

(1) Establishing and managing the KSC Apollo News Center at Cape Canaveral.

(2) Scheduling briefings and press conferences at KSC.

(3) Preparing for and managing operations at LC 39 Press Site.

(4) Providing countdown commentary as directed by the Apollo Information Director.

e. Public Affairs Officer, MSFC (Bart J. Slattery), designated Deputy Apollo Information Director (MSFC), will be responsible for the following:

(1) Directing all Apollo information activities conducted at MSFC.

(2) Providing information assistance at the KSC and MSC News Centers.

(3) Directing the European News Center, USIS, Paris.

f. Public Affairs Officer, GSFC (Ed Mason), designated Information Coordinator for Manned Space Flight Network.

g. NASA Headquarters Director of Special Events (Wade St. Clair), designated Coordinator for Protocol and Special Activities.

h. NASA Headquarters Chief of Public Information Audio-Visual Branch (Les Gaver), designated Coordinator for all Audio Visual activities during the Apollo mission.

i. KSC Public Information Officer (Gatha F. Cottee), designated Coordinator for Communications relating to the public information portion of the mission to include communications links between KSC and MSC, as well as special communications which may be needed before, during or after the mission.

j. MSC Public Information Officer (Bennett James) designated Senior NASA Information Representative onboard prime recovery ship. He will be responsible to the Apollo Information Director for NASA information activities and news pool operations in the prime recovery area.

4

k. The Director of Information, Air Force Eastern Test Range, (Col. James Smith), who is DOD Manned Space Flight Public Affairs Officer, is designated Assistant for Public Affairs, DOD Operations, for this mission.

Headquarters Public Affairs personnel will be flying from KSC to MSC in OMSF planes as follows:

Mission Director's T-1 day plane: Alibrando, Mittauer.

T-day plane: Scheer, Atchison, Allaway.

Other personnel assignments are contained in another section of this plan.

SECURITY

The Apollo Program Security Guide will be used to determine security classification of technical information related to the Apollo missions.

INFORMATION PROCEDURES AND OPERATIONS

The National Space Act of 1958 requires NASA to report its accomplishments fully, candidly and promptly. Information personnel involved in Apollo missions must be aware of this responsibility at all times in performing their official duties.

Responses to all queries on predicted time and location of reentry of launch vehicle stages or other artifacts of a Saturn/Apollo mission should be obtained from the Office of Public Information or the Public Affairs Office, OMSF, NASA Headquarters. This also applies to reported reentries or impacts of stages, fragments or other debris.

All information activities involving the Apollo Instrumentation Ships will be approved by the Apollo Information Director and coordinated with the Assistant for Public Affairs, DOD Operations.

All requests for news media interviews with OMSF Center Directors and the Apollo 11 mission officials must be cleared with the Apollo Information Director.

Apollo 11 accreditation of both newsmen and contractor representatives operating at KSC and/or MSC will be coordinated by NASA Headquarters Public Information Division. Contractor accreditation for KSC will be allocated by NASA, as outlined in the Contractor Activities section of this plan, because of facility limitations at LC 39 Press Site.

The official NASA headquarters Public Affairs Plan *for the Apollo 11 mission spells out in detail all of the department's operational procedures, with contributions from all fourteen NASA Public Affairs offices across the country. In a post-Paul Haney world, Julian Scheer exerted more control over the activities of each of the centers; Brian Duff, who now headed Public Affairs in Houston, rotated responsibilities of his staff in order to not create any more "celebrity PAO" staffers. As Scheer points out in the opening memo of his plan, while it is "not intended as a comprehensive document reflecting all of the four divisions of this office (Public Information, Special Events, Educational Programs, Media Development)," it does spell out major policies, which "reflect the open program conducted by this Agency. This is not subject to change under any circumstance." The document was published in early July 1969, prior to the Apollo 11 mission.*

Wernher von Braun and President Eisenhower. *Von Braun, Marshall Space Flight Center's first director, points out details on a Saturn rocket to President Dwight D. Eisenhower at the center's dedication ceremony, September 8, 1960. Two years earlier, on July 29, 1958, Eisenhower signed into law the National Aeronautics and Space Act, the federal statute that created NASA.*

Note, at the president's right, the double-panel display "The Team Behind the Saturn" on which the names and logos of the various contractors are showcased.

An Unprecedented Public Relations Partnership: NASA and Industry

"We sure didn't do the PR job by ourselves."

—Chuck Biggs, NASA Public Affairs Officer

As *Eagle* emerged from the far side of the Moon on July 20, 1969, Neil Armstrong and Buzz Aldrin were preparing to land their Apollo 11 lunar module in about a half hour—if all proceeded according to the mission plan. In Houston, flight controllers worked in the Mission Operations Control Room while VIPs watched from theater seating above. The ultimate goal of the Apollo program, initiated by President Kennedy a little more than eight years earlier, was so near that everyone in the room could feel the collective nervous energy.

Across the hall was the Spacecraft Analysis room, more familiarly known by its acronym SPAN. Here, representatives from the important contractors that built mission-critical components gathered for the impending landing. Engineers from all the leading contractors were available to lend their knowledge and expertise in the event of a problem with the spacecraft systems. During the years leading up to Apollo 11, contractor employees trained alongside Mission Control officers as members of a finely tuned partnership between NASA and civilian industry. Indeed, the fact that one's paycheck was issued by NASA or came from a private company made no difference when an issue arose; they worked together as a team to solve a problem.[1]

Simultaneously, at the nearby Nassau Bay Hotel, a more informal—yet no less emotionally engaged—assembly was gathering. Public relations professionals employed by contractors such as Boeing, Grumman, RCA, Honeywell, IBM, and Raytheon mixed with NASA Public Affairs people and reporters from around the world at the Escape Velocity Press Club. Inside, the large room was crowded and the atmosphere tense as hundreds of professional communicators listened to the astronauts' voices from the Moon.

During the preceding decade, the unprecedented cooperation between the government and the private sector was a decisive factor leading up to the goal achieved that day. In the eleven years since NASA was established, this complex relationship had grown and matured. And at its most troubled moment, the deadly Apollo 1 launch-pad fire two-and-a-half years earlier, the agency and its contractors confronted, and worked in unison to overcome, a crisis that significantly redefined their commitments and responsibilities to each other.

The United States government had existing ties to most of the major aerospace contractors well before NASA was created from the old National Advisory Committee for Aeronautics (NACA) in 1958. Businesses that had previously worked with the Department of Defense easily transitioned their sophisticated knowledge, technology, and manufacturing skills to help achieve America's emerging goals in space. And since NASA contracts followed guidelines similar to those already in place for defense purchases, such firms were well acquainted with the established business model. Most aerospace and defense contractors of the period worked on cost-reimbursable contracts, often in facilities owned by the government or rented from third parties. Within this paradigm, the government reimbursed allowable costs: labor and materials (direct costs) as well as overhead and administrative expenses (indirect costs). The government kept a watchful eye on all indirect costs and audited each major contractor at regular intervals. While the profit margins were relatively low, as the result of restrictions on expenses, the return on capital invested by the contractors was guaranteed. It was good for cash flow, if not for profits.

Significantly, such government contracts did not deem advertising or public relations an allowable overhead cost, so any

1. Thomas J. Kelly, *Moon Lander: How We Developed the Apollo Lunar Module*. Washington, D.C.: Smithsonian Books, 2009.

FACTS
& FIGURES

MANPOWER

Federal Employees	2,788
Vehicle Stage Contractors	6,466
Spacecraft Contractors	2,600
Unmanned Launch Contractors	1,495
Support Contractor Personnel	9,918
	23,267

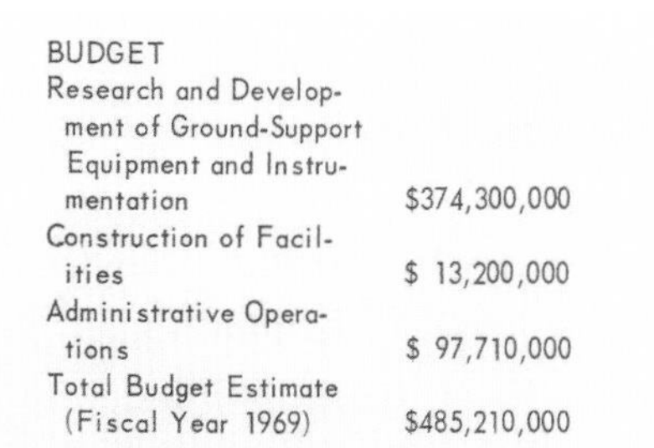

BUDGET

Research and Development of Ground-Support Equipment and Instrumentation	$374,300,000
Construction of Facilities	$ 13,200,000
Administrative Operations	$ 97,710,000
Total Budget Estimate (Fiscal Year 1969)	$485,210,000

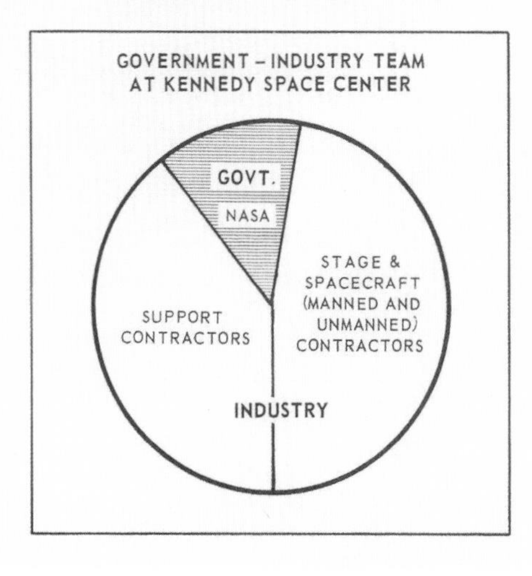

This pie chart from the NASA brochure "America's Spaceport," published in 1968, gives a striking visual picture of the size of the workforce at the Kennedy Space Center that was employed by NASA and the federal government compared to the private industry's workforce.

marketing dollars had to be drawn from the relatively small profit margins. Additionally, as was the case with defense contractors, press releases and advertising directly relating to NASA contracts had to be submitted for Agency approval in advance of publication. For the most part, prime contractors were discouraged from touting particular products in advertising and press releases. Instead, general corporate branding aimed at government customers and the investor community was the rule. Nevertheless, there were ways to work within the system to achieve mutual goals while enhancing the contractor's brand. For instance, when NASA sponsored an educational symposium for the press, contractors were often directly involved in the preparation and creation of presentation materials, an allowable expense under the published Federal Acquisition Regulation.

At its height in the late 1960s, the Apollo program funded employment for some 400,000 people and relied on 20,000 private-sector corporations. It also drew upon the work of many more at scores of universities. That the entire project was visible to the public via the world's news media—while it was taking place—made it all the more remarkable.[2]

The Escape Velocity Press Club and its counterpart, the Joint Industry Press Center (JIPC), also located in Houston, were run and funded by the contractors. JIPC was a place where reporters could pick up press materials and connect with manufacturers' PR representatives. Like the Escape Velocity Press Club and its often-boozy atmosphere, JIPC provided a place to socialize over a cup of coffee and refreshments. At the Cape launch site, the Apollo Contractors Information Center, similar to Houston's JIPC, was a contractor-run location where media could gather.

NASA had a lean public affairs team for such a large government agency; in July 1969, it consisted of only 146 employees in fifteen locations across the country.[3] Therefore, the contractors' public relations representatives became primary sources of information about technical aspects of the Apollo missions for thousands of journalists.

"We knew each other, we talked, and as the program gained momentum in '68 and early '69, all the PR guys got together and tried to figure out how we could carve out a little bit for ourselves and be of assistance to the thousands of journalists around the world who were following the program and coming to Houston for the Moon landing," recalled Harold Carr, a public relations representative for Boeing at the time. "So we created the Joint Industry Press Center and we ran that for all of the Moon flights. It was a 24-hour operation. We leased space in one of the motels there on the strip, served coffee and food, and set up tables and chairs—everything we could do to handle the overflow of journalists."[4]

Whenever the press had detailed questions about spacecraft design, hardware, or spaceflight subsystems, the answers came from PR representatives from the companies that built the components. Chuck Biggs, a NASA Public Affairs officer during Apollo, remarked years later, "We sure didn't do the PR job by ourselves. We needed representatives from Rockwell, Martin Marietta, and all the other contractors to do the job. By head count, we had more contractors' public relations people than we had NASA employees."[5] It was a true communications partnership created to tell the Apollo story worldwide through the media.

"For every launch at the Cape, we had a technical person on hand, available to the press," recalled Carr. "If anything were to happen during the launch sequence, or when Boeing's first

2. The unprecedented cooperation between contractors and NASA is explored in detail in *The Apollo Spacecraft: A Chronology* (NASA SP-4009), a four-volume set of books by Ivan D. Ertell, Mary Louise Morse, Jean Kernahan Bays, Courtney G. Brooks, and Roland W. Newkirk, published 1972–1978 by National Aeronautics and Space Administration, Scientific and Technical Information Office.

(notes continued opposite page)

stage of the Saturn V—the S-1C—was doing its thing, we were prepared to speak in detail about what was going on." Carr and his associates came to each launch equipped with a suitcase full of documents and they made themselves readily available at the Cape Kennedy press center.

The July 1969 mission of Apollo 11 attracted hundreds of novice and less-experienced reporters from around the world. However, the members of the media who had previously covered the Gemini program and the early Apollo missions were highly conversant about space travel and the science and technology that lay behind it. "A large segment of the media was very knowledgeable and intelligent at that time," recalled Doug Ward, an Apollo-era NASA Public Affairs veteran. "That was a plum assignment for science and technology writers and commentators. Many of them had followed the program from Gemini; some of them went back all the way to Mercury, in 1961. These reporters didn't require an awful lot of care and feeding, but they had a voracious appetite for information."[6]

Seeking Recognition and Advantage

Naturally, every Apollo contractor hoped their PR efforts would result in positive coverage about their company in the thousands of news stories about the lunar landing. Nearly all were actively engaged in efforts to secure additional government contracts, so favorable press coverage and association with the Apollo program set them apart and gave them an advantage. To these ends, contractors developed elaborate press kits containing detailed information about each company's role in the program. These printed materials supplemented the in-person expertise available to reporters during the course of the missions.

Guaranteeing that a kit of materials would stand out from the crowd was no easy task. A mere printed release together with press-ready photos tucked inside a pre-printed folder wasn't enough when hundreds of other public relations people were also courting the media and handing out documents. But a creatively produced press kit with well-presented, clear, informative materials often captured attention and led directly to mentions in the media.

Hasselblad, the Swedish camera manufacturer, chose to showcase what their product did. In a thick press kit sent to the media and issued at Apollo launches from Apollo 12

These photos, taken on the Apollo 9 mission, were included in a 36-page booklet as part of a press kit issued by Hasselblad, the Swedish camera manufacturer. The booklet showcased to the media how the cameras were used on missions. This was a rare instance in which NASA permitted mission images with products being used to be reproduced for commercial purposes. The upper photo shows astronaut Rusty Schweickart with the Hasselblad camera as he tested the new Apollo A7L spacesuit, the first spacesuit to have its own life-support system, as would be required on the Moon. The lower photo shows astronaut Dave Scott emerging out of the hatch, camera in hand, taking Schweickart's picture as Schweickart takes his.

3. In 1969, the Manned Spacecraft Center in Houston employed thirty-one people in the Public Affairs Office. This included news services, education, manned flight awareness, the audio-visual department, and others. The number of people dedicated to serving journalists during missions was fewer than twenty. At NASA Headquarters in Washington, D.C., there were forty-nine Public Affairs staffers, much of whose work was government and contractor relations. At Kennedy Space Center there were fifteen Public Affairs employees. (Source: *Apollo 11 Public Affairs Plan*, July, 1969.)
4. Harold Carr, interview with the authors, December 14, 2011.
5. Chuck Biggs, interview with the authors, November 10, 2011.
6. Doug Ward, interview with the authors, November 29, 2011, and December 12, 2011.

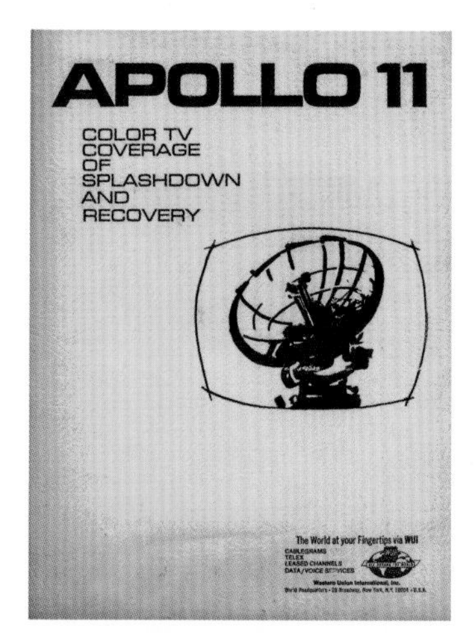

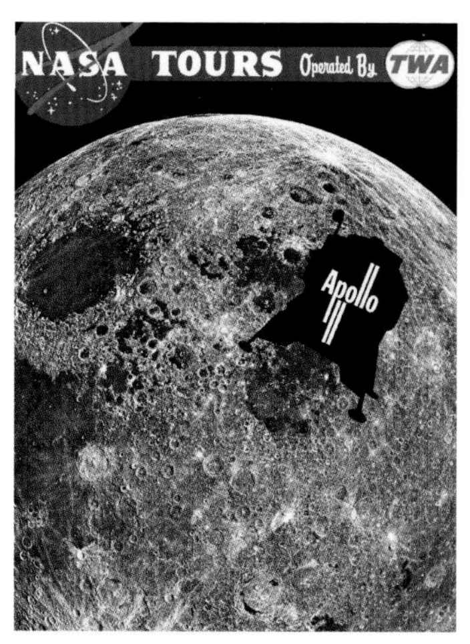

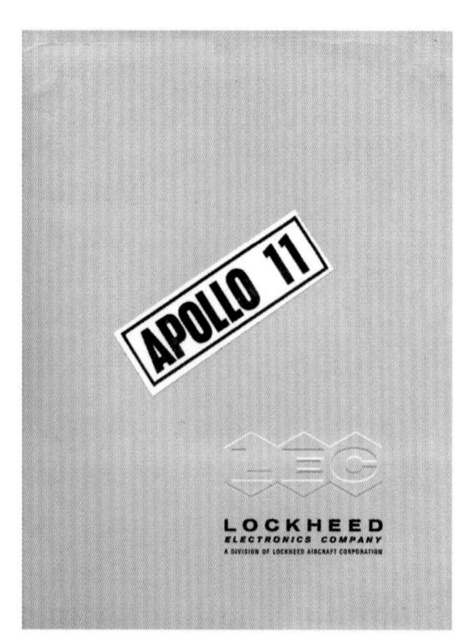

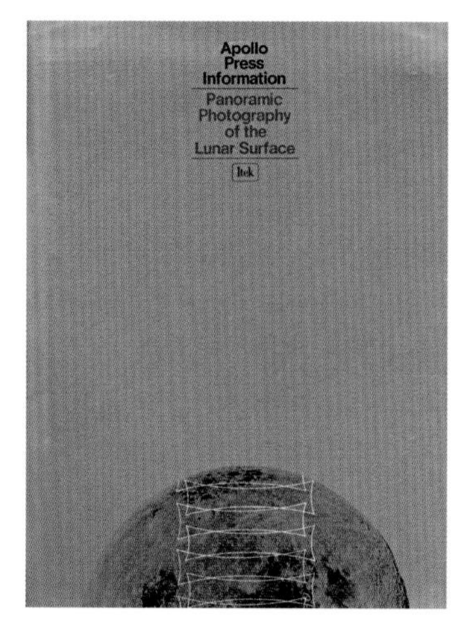

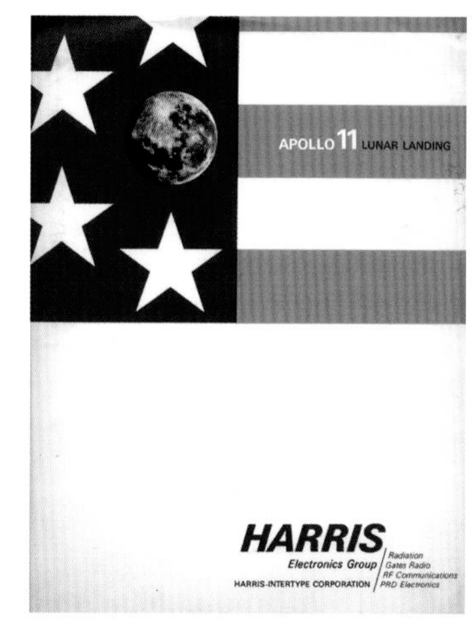

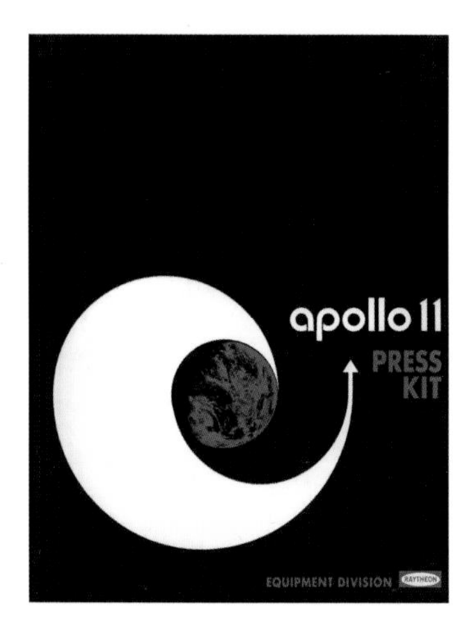

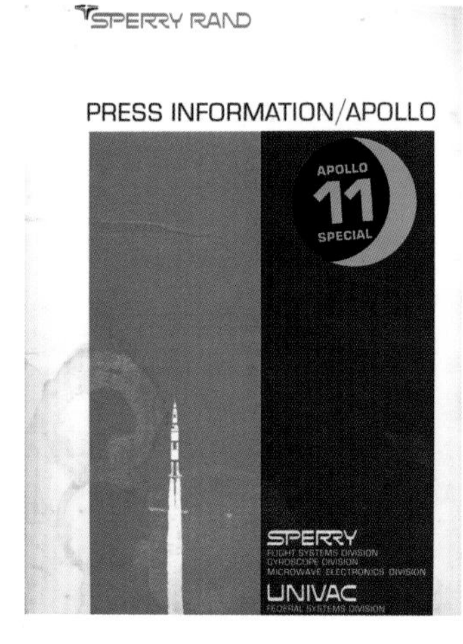

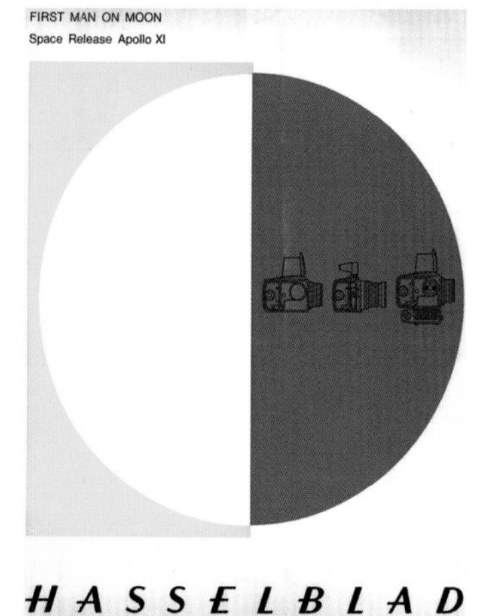

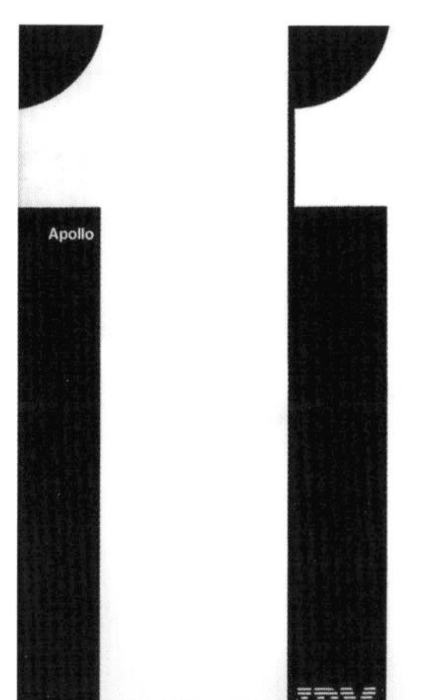

Apollo 11 Press Kit Covers. *Press kits prepared by the public relations staffs at the major contractors for the Apollo 11 mission provided valuable additional information not found in NASA-issued releases. Reporters and editors working on stories about the lunar landings had access to such documents from more than one hundred manufacturers.*

onwards, the single press release headlined "First Camera on Moon" seems like an afterthought next to the stunning 36-page book of photographs included within the kit. The book, with an introduction by Victor Hasselblad, features dozens of photographs from the first flights of the lunar module on Apollo 9 and 10 and the first lunar landing of Apollo 11. It demonstrated to reporters in stunning detail precisely what a Hasselblad camera could do and why it was chosen by NASA to document the Apollo missions. The kit also included seven lunar surface photographs ready to be printed in magazines and newspapers. Each was supplemented by a caption detailing camera information such as "Photo taken with the Hasselblad Electric Data Camera with reseau plate, Zeiss Biogon f. 5,6/60 mm lens, Synchro-Compur shutter. The camera was fitted with a specially-designed Hasselblad 70mm magazine for use on the Moon."

Clearly, if a product was as simple to understand as a camera used on the lunar surface, the manufacturer had an edge getting it mentioned in the media. But most companies weren't in such a position. For spacecraft hardware manufacturers, capturing the media's attention required a bit more creativity. Realizing that reporters would need to interpret complex data for readers and viewers in accessible language, several contractors included visual representations of esoteric technical information in their press kits. For example, Grumman included a multi-sheet graphic that dissected their lunar module. A series of cutaways printed on layers of clear acetate sequentially revealed the detailed inner workings of the spacecraft, providing a comprehensible visualization of the LM's construction at the Grumman plant.

General Motors's AC Electronics, a company that manufactured guidance and navigation equipment for Apollo, created a mission plan that mapped the major events during a flight so that reporters could visualize what was happening at each step. TRW, Raytheon, and North American Aviation went even farther, creating slide rule-style analyzers. North American Rockwell's PR department issued an "Apollo Mileage and Speed Converter" that reporters could use to convert the reported data, such as feet-per-second, into statute miles-per-hour, something laymen could more easily understand. Raytheon's "Mission Analyzer" was used to convert NASA's Ground Elapsed Time into a more familiar time and day of the

__Apollo 13 Mission Analyzer.__ Raytheon produced the circular "Apollo 13 Mission Guidance and Navigation Analyzer," and similar ones for other Apollo missions, as an aid to understanding the important activities during the missions. Of course, this analyzer reflected the mission plan had everything proceeded normally. When the oxygen tank exploded in Apollo 13's service module, the lunar landing was aborted and the mission became one of survival.

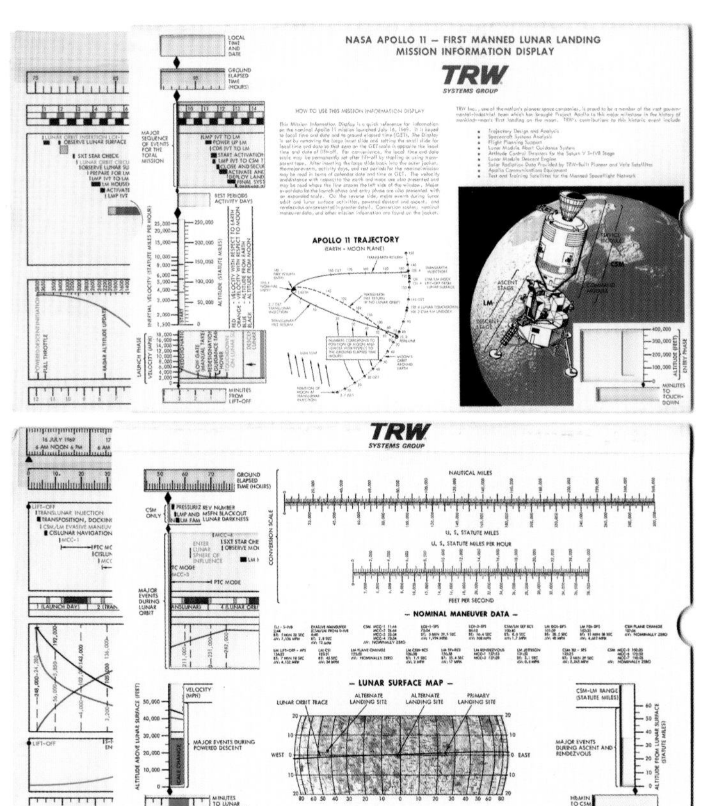

__TRW Mission Information Display for Apollo 11.__ TRW manufactured the lunar module descent engine and other components for Apollo 11, as well as the trajectory design and analysis. They made this two-sided slide rule as a giveaway to reporters and dignitaries. It is, itself, an impressive feat of information design.

Press materials, left column: *The Apollo Contractors Information Center was, like the Joint Industry Press Center in Houston, a place where reporters could gather to pick up press materials and information from the contractors and major subcontractors. It was entirely contractor-funded and operated in the city of Cape Canaveral.*
Center column: *In a photo dated July 25, 1969, an engineer at the Instrumentation Laboratory of the Massachusetts Institute of Technology, a major NASA contractor, is shown checking on-board flight guidance computer programs for the Apollo Command Module using the M.I.T. Command Module Simulator. Below is Sperry Rand's UNIVAC computer system; at bottom is the 15-foot, gyro-stabilized parabolic antenna built by Western Union for its portable earth station aboard the USS* Hornet.
Right column: *a Mission Control station built by Philco; a guide to solid fuel modules manufactured by Morton-Thiokol; the capsule recovery logistical plan by Brown & Root; IBM's mission-tracking wheel, featuring the principal systems components built by the company.*

APOLLO CONTRACTORS INFORMATION CENTER

COLOR TV, LOUNGE AREAS	TELEPHONE CALL BOARD
PRESS WORKING AREA	TELEPHONE SERVICE
CONTRACTOR NEWS RELEASES	ACIC REPS. STAFFING
AIRLINE SERVICES	AIR EXPRESS

ACIC on second floor, Apollo News Center
Cape Canaveral, Fla. 784-1260

ACIC Apollo 11 Members

Prime Contractors	Company Representative	Telephone
Bendix	Bill Lyerly	867-8083
Boeing	John Payton	784-3122
Catalytic-Dow	Carl Weisiger	269-5500
Chrysler	Tom McDonald	784-6477
Federal Electric	Kurt Voss	784-1303
General Electric	Gene Folkman	783-0506
Grumman	John Vandegrift	784-1718
IBM	Leon Bill	784-9782
McDonnell Douglas	Ken Grine	269-4100
North American Rockwell Space Division	Joe Cullinane	783-9005
North American Rockwell Rocketdyne	John Ulf	783-9005
Pan American	George Dill	494-5784
TWA/KSC base support	Tom Winfield	632-7400/267-5377
TWA/NASA Tours	George Meguiar	867-2050/269-3366

Subcontractors

American Standard	Philco-Ford Corp.
Collins Radio Co.	Radiation
General Dynamics	Ryan Aeronautical Co.
Hamilton Standard	Sperry-Rand
Honeywell, Inc.	Thiokol
Johns-Manville Corp.	TRW Systems, Inc.
Lockheed Propulsion Co.	Westinghouse Electric Corp.
Pratt Whitney Aircraft Division of United Aircraft Corp.	

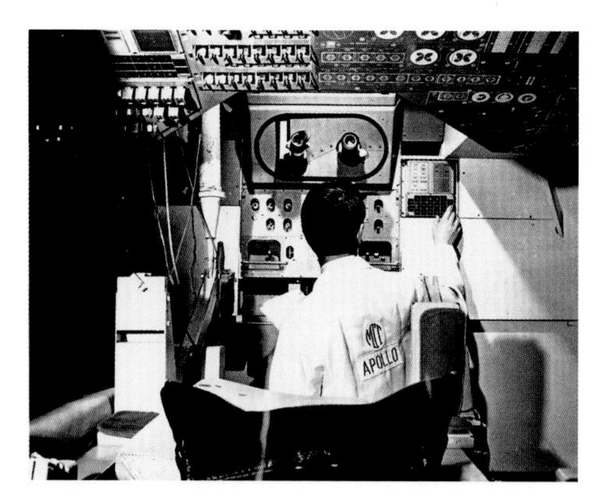

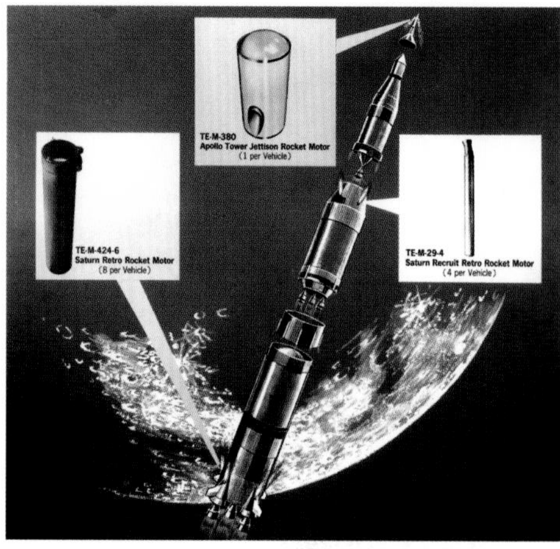

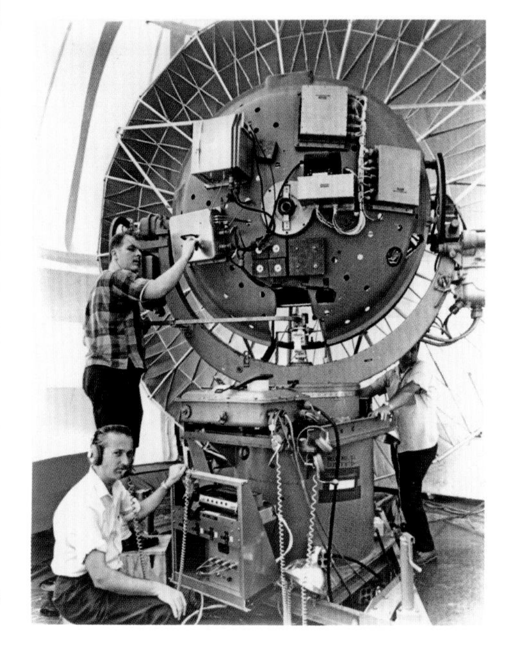

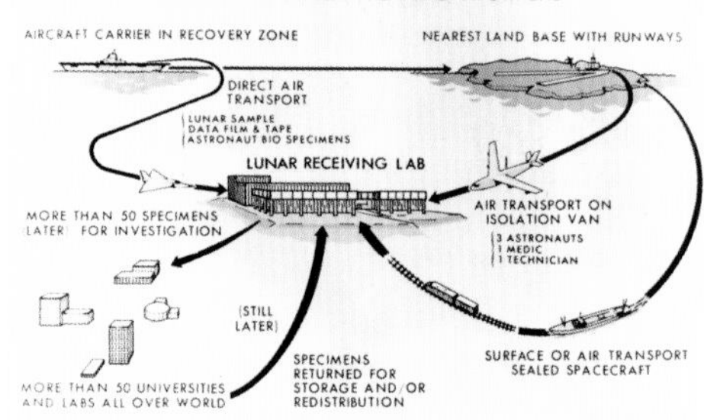

APOLLO 11 SIDEBAR NEWS

<u>IMMEDIATE RELEASE</u>

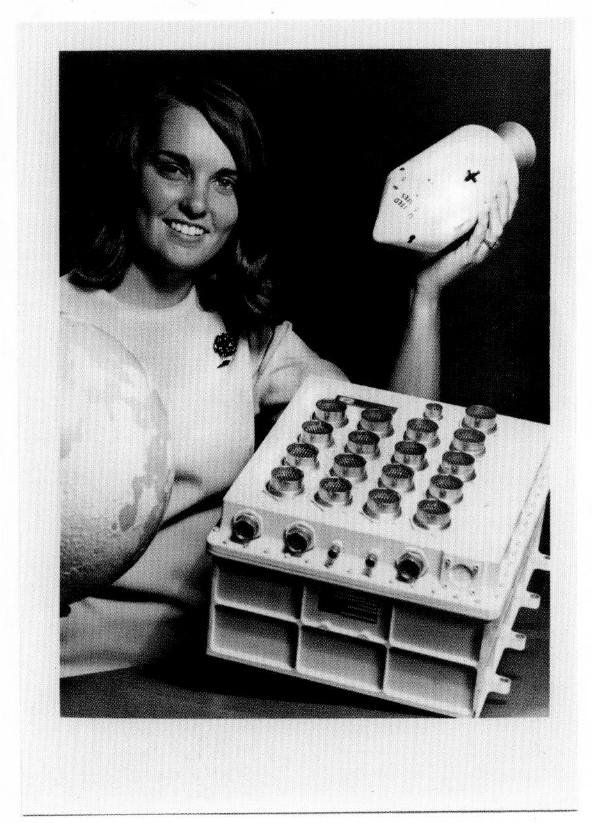

<u>PRETTY JANET HINKSON</u>

poses with what will be the central nervous system of the Apollo 11 Command Module. The system, a precision telemetry unit, collects data from tiny sensors, then codes and condenses the data for radio transmission back to earth. The information collected includes everything from the speed of the spacecraft to the space traveler's heart rate. Over 21,000 electronic components are packed into this pressurized unit, the size of a hatbox. It was designed and developed by the Electronics Group of Harris-Intertype Corporation.

-30-

HARRIS
Electronics Group | Radiation · Gates Radio · RF Communications · PRD Electronics
HARRIS-INTERTYPE CORPORATION · 55 Public Sq. · Cleveland, Ohio 44113 · (216) 861-7900

SPACE AGE needleworker "weaves" core rope memory for guidance computers used in Apollo missions. Memory module will permanently store mission profile data on which critical maneuvers in space are based. Core rope memories are fabricated by passing needle-like, hollow rod containing a length of fine wire through cores in the module frame. Module frame is moved automatically by computer controlled machinery to position proper cores for weaving operation. Apollo guidance computer and associated display keyboard are produced at Raytheon Company plant in Waltham, Massachusetts.

SPACE AGE 49925A

Public Relations
Raytheon Company
Lexington, Massachusetts
(617) 862-6600 ext.415

Press kit materials—the rare presence of women. *The space program of the 1960s was dominated by white men; women were most often seen, when they were seen at all, in secretarial roles behind the scenes or as product models, as in the Harris Electronics Group's press release (above) featuring "Pretty Janet Hinkson" with a critical, though hardly photogenic, component from Apollo 11 command module telemetry system. At Raytheon (right), a woman is building a memory module for the Apollo guidance computers.*

Apollo Spacecraft News Reference. *Even more elaborate than a basic press kit, the* Apollo Spacecraft News Reference *books, created for journalists, were presented in three-ring binders containing hundreds of tabbed pages of details about the space vehicles and subsystems. They included photographs, drawings, and charts. Interestingly, the credits on the books' covers indicate the extraordinary public relations partnership between NASA and the contractors who built the spacecraft. This volume, describing the lunar module, was produced by Grumman Aircraft Engineering Corporation in collaboration with NASA's Manned Spacecraft Center. The command module book was produced by NASA MSC and North American Aviation's Space and Information Systems Division.*

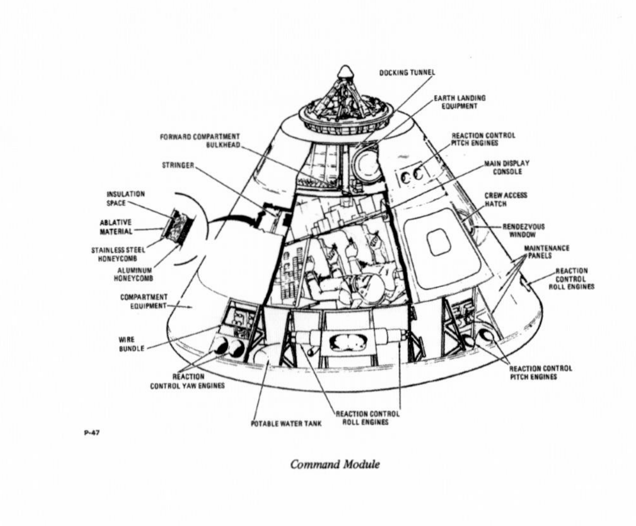

APOLLO NEWS REFERENCE

DESCENT STAGE

The descent stage is the unmanned portion of the LM; it represents approximately two-thirds of the weight of the LM at the earth-launch phase. In addition to containing the descent propulsion section, the descent stage is designed to:

Support the ascent stage

Provide storage to support the scientific and communications equipment used on the lunar surface

Provide for attachment of the landing gear

Serve as the ascent stage launching platform

The descent stage is separated into five equally sized compartments that contain descent propulsion section components. The center compartment houses the descent engine; fuel, oxidizer, and water tanks are distributed in the remaining four compartments.

COMMAND MODULE

Dimensions

Height	10 ft 7 in.
Diameter	12 ft 10 in.
Weight (including crew)	13,000 lb
Weight (splashdown)	11,700 lb

Command Module

AS-2

Grumman

week standard. Among the most elaborate was TRW's "Mission Information Display" which offered a variation on Raytheon's calculator, with the addition of an estimated time to touchdown for the lunar landing based on altitude as well as a listing of major event data. The public relations staff that instigated these unusual and valuable tools were not only creating good will for their corporate employers, they were also hoping that these objects might inspire a reporter somewhere to mention their company in print, while citing their connection to the Apollo program.

During the 1960s, it was common practice for public relations departments to catch the attention of wandering eyes with product photos featuring a comely female face or an attractive figure. Mirroring the macho world of test pilots, the communications industry was largely male and highly competitive. Contractors went so far as to produce news releases touting the merits of their pretty employees, some even shown fondling long cylindrical objects in glossy photos.

Of all the press packages produced during the Apollo years, the *Apollo Spacecraft News Reference* binders were in a league of their own. These highly illustrated manuals were jointly produced by NASA and the spacecraft's prime contractors, Grumman Aircraft Engineering Corporation and North American Rockwell. Each elaborate three-ring binder was filled with more than 100 pages of technical data, charts, diagrams, and specifications, rapidly establishing itself as the primary technical encyclopedia for any journalist covering the Apollo program. They were created partly in response to a barrage of press inquiries during the months leading up to the first missions. "Reporters kept calling with one question, and then

APOLLO NEWS REFERENCE

Function

The service module contains the main spacecraft propulsion system and supplies most of the spacecraft's consumables (oxygen, water, propellant, hydrogen). It is not manned. The service module remains attached to the command module until just before entry, when it is jettisoned and is destroyed during entry.

Major Subsystems

Electrical power
Environmental control
Reaction control
Service propulsion
Telecommunications

The service module is a cylindrical structure which serves as a storehouse of critical subsystems and supplies for almost the entire lunar mission. It is attached to the command module from launch until just before earth atmosphere entry.

The service module contains the spacecraft's main propulsion engine, which is used to brake the spacecraft and put it into orbit around the moon and to send it on the homeward journey from the moon. The engine also is used to correct the spacecraft's course on both the trips to and from the moon.

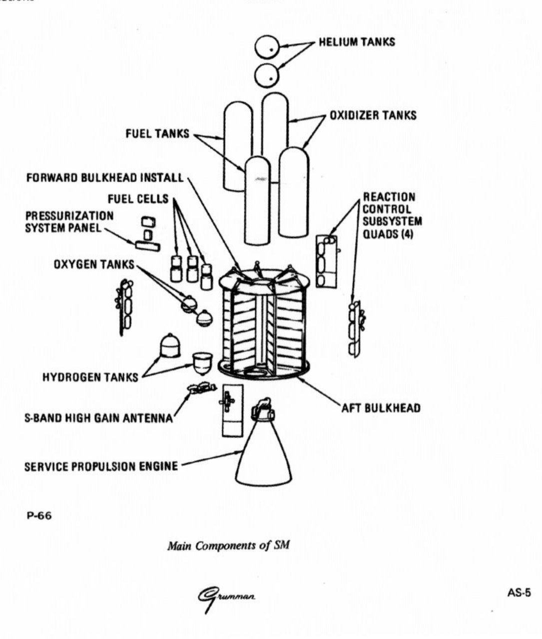

Main Components of SM

Grumman AS-5

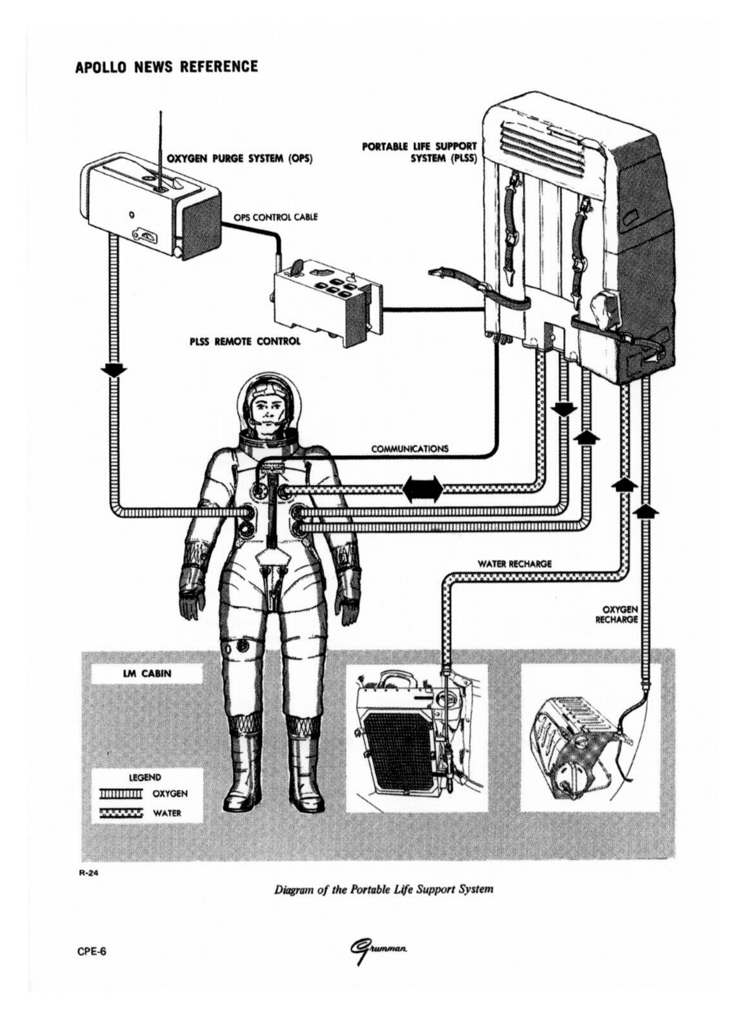

Diagram of the Portable Life Support System

the next day they would have another question," recalled Bob Button, a journalist and veteran of Grumman's public relations department. (He had earlier served in NASA's Public Affairs Office). "I went to the president of the company and proposed that we conduct an educational class; invite all the press to this one big classroom where we would have our experts available. But we also needed to have a handout."[7]

North American Rockwell had just produced their impressive *Apollo Spacecraft News Reference* manual covering the command and service modules, and thus it served as the ideal template for NASA and Grumman's "classroom handout." But the job of putting it together was unlike anything that Grumman had ever done, and it had to be written and published within two months to be ready by January 1969. The man assigned to oversee the project was Dick Dunne, Grumman's Director of Public Affairs for Space and its chief spokesman. A veteran technical writer who had previously worked on Grumman's swept-wing Navy fighters, Dunne supervised the writing and editing of the lunar module reference manual, sometimes coordinating with consultants from firms like Hamilton Standard to produce special sections within book. For the section on the Moon, Dunne contracted Richard Hoagland, the former Curator of Astronomy at the Springfield (Mass.) Science Museum. The first edition of the lunar module *Apollo Spacecraft News Reference* was printed in an edition of 2000 copies in time for the first Grumman class in early 1969.[8]

Journalists and network correspondents were invited to attend the two-day symposium held at the auditorium at Grumman headquarters in Bethpage, New York. (Later sessions were held at other locations.) "Everybody came to that thing,"

7. Bob Button, interview with the authors, March 13, 2012.
8. Dick Dunne, interview with the authors, May 2, 2013.

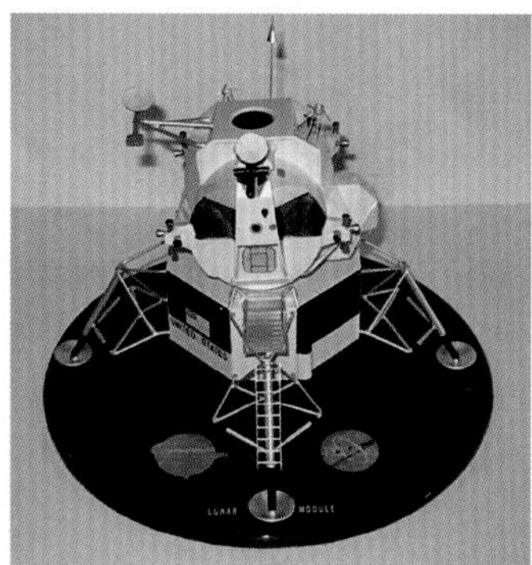

Clockwise from top:
***CBS Evening News* anchor Walter Cronkite** *holds the Grumman lunar module model. Cronkite and his counterparts on NBC and ABC frequently used spacecraft models during their extensive broadcast coverage of the Apollo lunar missions.*

A ***scale model of the LM*** *of the type offered to television journalists for use on air. Being the first craft used by humans to operate in an environment without an atmosphere, the lunar module's strange design required a visual explanation to viewers. Its manufacturer, Grumman Aerospace Corporation, provided models to politicians, dignitaries, subcontractors, astronauts, and flight directors. Such models were often used as props in NASA's photographic portraits of the astronauts. They also graced the desks of Grumman staff, and during the Apollo era they could be purchased for $29.95 in the Grumman employee shop.*

The ***Duro Marker****, a lightweight, aluminum felt-tip pen, was selected by NASA to be flown on all Apollo missions. In total, twenty-seven Duro Marker pens were flown to the moon, and twelve of those landed on the lunar surface. The Duro Pen Company of Brooklyn, New York, sold a similar pen to consumers in packaging that celebrated the marker's role in the lunar missions.*

The off-the-shelf ***Exer-Genie*** *personal exerciser was perfect for spacecraft use because of its small size. "We were looking for something to exercise with, so we told our training team to propose something," said Walt Cunningham, lunar module pilot on Apollo 7. Later, the product was renamed the "Apollo Exerciser" and sold to consumers in packaging like this.*

Button recounted. "It was like a school, with us anticipating questions and collecting others that we had already received. We brought in speakers: the lunar module project manager addressed the class, as did trajectories experts, who explained how the lunar module actually lands on the Moon." In an interview conducted two decades later, Walter Cronkite recalled how, during the months before the lunar landing, he strenuously studied these manuals in detail: "I had to start from scratch because I am not mechanically trained in any way, certainly not scientifically. . . . I took NASA's manuals and books, and I did my homework. I studied like fury." Cronkite's son Chip particularly remembered how his father, in early 1969, poured over those huge binders like a middle-aged graduate student who had just returned to school.[9]

After the Grumman symposium, the lunar module edition of the *Apollo Spacecraft News Reference* became the journalists' bible when covering the moon landings. Inevitably, as additional members of the press and others became aware of its existence, Grumman was flooded with requests for additional copies. Quantities were limited, however, and ordering a reprint would be prohibitively expensive, so to ensure fair distribution to as many different news organizations as possible, Grumman's public affairs office kept a working list of the names and affiliations of everyone who received a complementary copy, with an understanding that they were to be shared within a particular news office. (Following the Apollo 11 mission, Grumman and NASA produced two later, revised looseleaf editions, which described the heavier lunar modules used in the later flights.)[10]

In their hunger to tell a story visually, television journalists eagerly relied upon contractor-provided props and scale model spacecraft in the course of their on-air reporting. Faced with the task of explaining the basic operation of a spacecraft or a rocket engine to television viewers, reporters discovered that scale models were ideal. North American Rockwell, the prime contractor for the command module, and Grumman Aircraft Engineering Corporation, the prime contractor for the lunar module, supplied a number of models, in various sizes, of the spacecraft they built to all of the major networks. Parts and equipment manufacturers gave television reporters parachute material, switches, dials, samples of dried space food, and the pens used during the missions.

EVERYBODY WHO'S BEEN TO THE MOON IS EATING STOUFFER'S.

When the Apollo 11 astronauts got back to down-to-earth food, they got back to Stouffer's. For four good reasons: flavor, quality of ingredients, convenience and careful preparation.

Of all the foods on earth, Stouffer's Frozen Foods met or exceeded every NASA specification.

So, 14 Stouffer's main dishes, side dishes and meat pies are featured on the menu throughout the critical postlunar quarantine period.

One more interesting thing. Stouffer's made this food for the astronauts exactly as they make it for you. Nothing more. Nothing less.

For the people you love, Stouffer's plays it straight.

Stouffer's
Frozen Prepared Foods

STOUFFER FOODS

Companies of all kinds were eager to showcase their association with the Apollo program to the public, even for products as mundane as the food the astronauts ate on the mission. In these consumer advertisements intended for popular magazines, Stouffer's prepared foods and Del Monte dried fruits promoted their use on the missions. Del Monte's ad featured a drawing by well-known New Yorker *cartoonist George Price, a play on the twin improbabilities of the influence of dried fruit on a successful space mission and a still-suited, returning astronaut being interviewed on a city street. The Stouffer's ad assumes the appearance of an urgent announcement.*

NASA rules did not permit companies to show photos of astronauts eating in space and, importantly, the companies could not say that the astronauts used particular products. They could, however, say that the product (in this case food) was chosen for the mission and carried aboard the spacecraft. The creative staff at the advertising agencies faced the challenge of creating images and copy that implied the astronauts used a product, even though they were not permitted to say so directly.

Whenever a consumer product was used on an Apollo mission, its manufacturer usually created new packaging celebrating that fact. Advertising and packaging for General Foods's instant orange drink Tang, the Duro Marker, and the Exer-Genie personal exerciser emphasized that these products were selected by NASA for use in the American space program.

As the relationship between the contractors and the press was established to explain the technological aspects of the Apollo story, the business side of the story was often left uncovered, without journalistic scrutiny. Reflecting on the news coverage of the era, Mark Bloom, the *New York Daily News*'s lead writer covering the Apollo flights, suggested that it could have been better. "If we were to cover it today, we would have a business writer covering it as well. We'd have somebody covering what this all means to Grumman and Rockwell and Boeing."[11]

Communicating Directly with the Public

There was certainly no guarantee that a business's investment in public relations would yield positive press coverage. So to ensure that as many as possible were aware of their role in humankind's greatest adventure, some companies conducted consumer advertising campaigns highlighting their involvement. Eastern and United Airlines promoted the fact that they flew routes to cities frequently visited by business travelers working on Apollo. The food manufacturers Del Monte, Stouffer's, and General Foods described products consumed during the missions. And producers of goods used on Apollo missions that were also available to consumers—Omega watches and Sony tape recorders, to name just two—created advertising campaigns that linked their brand to the space program, underlining the reliability, durability, and craftsmanship that led NASA to choosing their products.

9. Walter Cronkite and Don Carleton, *Conversations with Cronkite*. Austin: University of Texas Press, 2010, p. 232; and Douglas Brinkley, *Cronkite*. New York: HarperCollins, 2012, p. 410.
10. Dick Dunne, interview with the authors, May 2, 2013.
11. Mark Bloom, interview with authors, January 13, 2012.

Tang, the powdered drink mix from General Foods, was closely associated with the space program as the result of heavy advertising. The print advertisements shown here were published during the Gemini and Apollo programs.

Tang also advertised on television. A spot from late 1960s opens with an image of a "dinner of the future," with foods contained in plastic pouches, like those used on Apollo missions, arranged in a neat row on the plate. "This is a typical meal served to astronauts aboard Apollo spaceflights: oatmeal, sausage, toast, applesauce. And, in a special zero-gravity pouch, Tang, the energy breakfast drink. Tang. With rich, natural flavor and more vitamin C than orange juice." Cut to a typical breakfast setting with a big jar of Tang: "But Tang's biggest role isn't in the NASA space program. It is right here on earth."

A Tang commercial from 1970 opened with a young boy in pajamas getting out of bed: "The next time you get up in the morning, make yourself an ice-cold glass of Tang." As he puts on his slippers and stirs some Tang in a glass, the image jumps back and forth between the boy in his pajamas and the boy in a spacesuit. "Tang is for earthmen who just don't want to be earthbound." The boy, wearing the spacesuit, leaves his home through the front door. "Drink Tang and go." He steps down a ladder on a lunar module-like spacecraft. "Tang. The energy drink for earthmen."

Not only did NASA allow manufacturers to promote their role in the lunar missions, it also granted (with few restrictions) the reproduction of government photographs in commercial advertising. "There was no copyright asserted for these pictures if they were used in advertising," explained Mike Gentry, a veteran of the photo section in NASA's Public Affairs Office. "But we did not want to imply an endorsement by NASA of a given product, process, or service."[12] NASA did, however, require the submission of advertising material for approval. This policy, established by Public Affairs chief Julian Scheer, was, on the whole, very accommodating. According to veteran Public Affairs officer Doug Ward, "It was written into the NASA contracts that suppliers would submit any advertising material for approval. If what they wanted to say in an ad was factually correct, we would not only let them go, we would encourage them to do it."[13] When reviewing the ad copy, NASA personnel sought to remove any language implying an endorsement, or a statement that a product was actually used in a specific manner. While it was fine to state, for example, that Omega watches traveled to the Moon, the manufacturer was not permitted to imply that their watch was used to time a particular event, nor could the copy suggest that a particular astronaut wore the timepiece.

The spectacular photographs shot in space were paid for with taxpayer dollars and, accordingly, they were freely available for use in advertising. (NASA preferred to be given a photo credit.) However, NASA attempted to be very careful about photographs showing a recognizable astronaut because the agency's standards of conduct did not permit NASA employees—astronauts included—to appear in advertising or commercial situations. This policy also made it impossible for companies to use photos showing a product actually being used. While NASA had standards of conduct in place for NASA employees, an entirely different situation existed when

12. Mike Gentry, interview with the authors, November 21, 2011.
13. Doug Ward, interview with the authors, November 29, 2011 and December 12, 2011.

How can a man in a $27,000 suit settle for a $235 watch?

The Apollo-Soyuz spacesuits, like those for every preceding space mission, were designed especially for the job. Not surprising either. You'd hardly expect to find the equipment for the flight through space to this historic America-Russia meeting ready-invented in the shops.

Yet that's how the astronauts found the Omega Speedmaster, their watch.

In 1965 NASA picked up a Speedmaster, as simply as you do in your local jewellery shop. And they made it standard flight equipment for every astronaut because, unlike any other chronograph tested, whatever NASA did to the Speedmaster, it stood up.

If you're wearing an Omega Speedmaster you can be proud of it – numerous space missions, six moon landings, and now, almost unbelievably, America and Russia together. For any other watch, the shock would be too much.

1. Omega Speedmaster Professional Chronograph. Standard issue to the American astronauts.
2. Omega Speedmaster 125. Officially certified automatic chronograph chronometer.
3. Omega Speedsonic f300. Officially certified electronic chronograph chronometer.

Ω OMEGA

Out of this world.

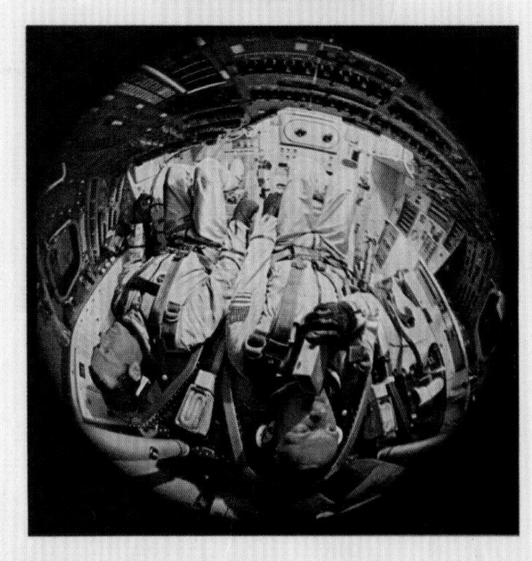

Sony/Superscope tape recorders

A tape recorder is an electronic log of man's most ambitious adventure.

It is the essence of human emotion. Or the fantasy of untold generations suddenly frozen into fact.

A tape recorder is philosophic reflections. Scientific observations. Or the innermost personal impressions of a lunar explorer. Preserved forever.

It is the commerce of everyday living. The earth song of a folk singer. Or the biology notes of a medical student.

Because your life is so full of sound, our life is building the finest tape recorders in the world. And the most popular.

This year, Sony offers more than 30 totally different models to choose from. Functional. Stereophonic. Portable. Deck. System. Reel-to-reel. Cartridge. And cassette.

How many ways are there to use a tape recorder? Use your imagination . . . then use Sony.

SONY SUPERSCOPE

You never heard it so good.

The Sony Model 50 pocket Cassette-Corder® illustrated above was selected by NASA for use in the Apollo Lunar Exploration Program. ©1969 Superscope, Inc., 8205 Vineland Ave., Sun Valley, Calif. 91352. Send for free catalog.

NASA photographs were created with taxpayer funding and, therefore, official NASA photographs could be used in commercial advertising with very few restrictions—and with no licensing fees. Their use, however, had to be approved by NASA Public Affairs staff, who made sure the copy was factually correct. Importantly, no advertising could imply an endorsement of a product or use language describing how, or if, the products were actually used. While it was fine to say, for example, that Sony/Superscope tape recorders were carried on Apollo missions, it wasn't permitted for the manufacturer to say which astronaut used the tape recorder or how.

NASA was very careful about photographs that showed a recognizable image of a particular astronaut, because the agency's standards of conduct did not permit NASA employees, including astronauts, to appear in ads or commercialize their role. This policy also made it impossible for companies to use photos showing a product actually being used. The Sony advertisement features professional models using the product in an Apollo command module mock-up.

employees left the agency. A number of astronauts became pitchmen for companies after leaving NASA.

The exchange of photographs was hardly one sided. Within the public relations and advertising departments of a number of contractors was a small staff dedicated to photographic services. Their many duties included creating images that documented the design and construction of their spacecraft components. As part of their partnership, NASA had free access to everything produced in conjunction with the contract—including any associated photography. Mike Gentry recalled that, "NASA considered the contractors' pictures part of the NASA collection. We didn't credit Boeing, for example, because it's a NASA paid-for picture."[14] Similarly, Grumman's photographs of the lunar module being built at their Long Island facility were created under a NASA contract and, accordingly, freely available to the space agency. NASA would use such photos for their own PR purposes without crediting Grumman.

Tang, the widely advertised powdered drink mix made by General Foods, benefited greatly from its association with the space program. The relationship was so important that executives assigned to the Tang account from the Young & Rubicam advertising agency would travel to Houston to review proposed advertising layouts with NASA. The agency executives wanted desperately to be able to say that the astronauts drank Tang, but Doug Ward recalled that, "We said, 'No. You can't say that.' We carry Tang and astronauts may or may not drink it, but they're not required to."[15]

Young & Rubicam representatives also requested the use of a photograph or video depicting an astronaut drinking Tang in space, but NASA wouldn't allow it. Nor would they consent to an image of an astronaut drinking an unrecognizable liquid. The space agency remained firm: the product was available to astronauts while in space, but it was their personal choice whether they drank it or not. It was left to the team at Young &

14. Mike Gentry, interview with the authors, November 21, 2011.
15. Doug Ward, interview with the authors, November 29, 2011 and December 12, 2011.

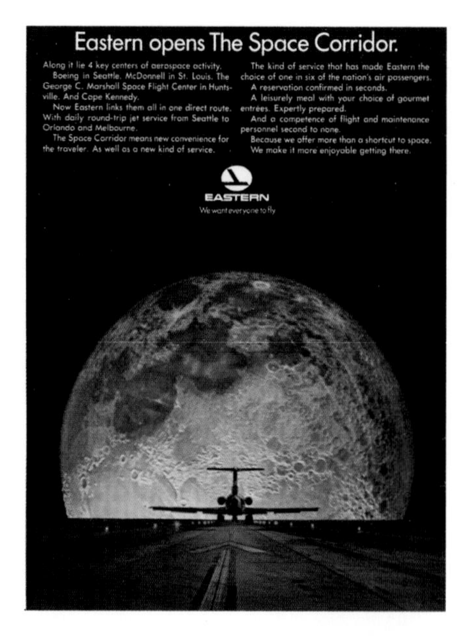

Many companies, both those directly involved with the space program, such as Ford and Boeing, as well as those not directly involved, such as Panasonic and Eastern Airlines, sought through their advertisements to associate their products and services with the leading edge exemplified by the program. The Allison and Boeing ads shown here were placed in Aviation Week & Space Technology, *and were directed toward the Department of Defense and their foreign counterparts.*

Rubicam to create an ad campaign emphasizing that American astronauts carried Tang on missions into space and use NASA photos to cement the link. Despite NASA's restrictions, Tang's ubiquitous television commercials and consumer ads are among the most remembered of the 1960s. So successful were the commercials, that there remains a popular misconception that Tang was developed for use in space. In truth, it was first introduced to the public in 1959 and experienced only mediocre sales. Only after it was flown on Mercury and Gemini missions, and after General Foods with Young & Rubicam created their marketing campaign, did consumer sales explode.

Trauma and Recovery: The TIE Program

All relationships are tempered and changed by times of crisis, and the unique partnership of NASA and its contractors was no exception. In early 1967, the Apollo program suffered its greatest tragedy, when the crew of the Apollo 1 spacecraft—Gus Grissom, Ed White and Roger Chaffee—were killed by a sudden and violent fire that erupted inside the sealed command module during a seemingly routine launch pad test. A combination of factors contributed to the accident, the most decisive being NASA's decision to use a pressurized 100% oxygen environment within the spacecraft. It took a single spark less than ten seconds to transform that dangerously volatile atmosphere into a raging inferno. (The spark is believed to have originated in a wire contained within the craft's urine collection system.)

Everyone involved in the program understood that the conquest of space was a dangerous enterprise, and that accidents were bound to occur. However, the stark realization that the Apollo fire had been preventable only intensified the depth of the despair and the resulting reaction. In the immediate aftermath, some even wondered whether the lunar missions might be canceled.

NASA quickly convened an accident review board. The agency also solicited suggestions from companies in the aerospace industry. Notable was the response from Boeing President William Allen: "We'll help the nation in any way NASA wants." NASA took Allen up on his offer and drafted a $73 million, one-year contract with Boeing to assist and support in the performance of specific technical integration and evaluation functions, a systems analysis that came to be known as Apollo TIE (Technical Integration and Evaluation). Boeing was well equipped for this assignment; it had experience coordinating geographically complex projects such as the Minuteman ballistic missile program. Allen assigned 2,000 managers to the TIE project who would scrutinize every drawing and design detail relating to the Apollo spacecraft in order to pinpoint any possible flaws that could have led to the accidental fire.

The Apollo TIE team uncovered communications deficiencies amongst the various parts of NASA, and they learned that some NASA centers were run with a great deal of latitude. "Houston didn't talk to the Cape, and the Cape didn't talk to Huntsville, and Huntsville didn't talk to Washington," recalled Harold Carr of Boeing, who worked with the Apollo TIE team. "I think the power at the NASA centers was ready to create a disaster someplace along the line, and the fire at the Cape shined a light on the problem."[16]

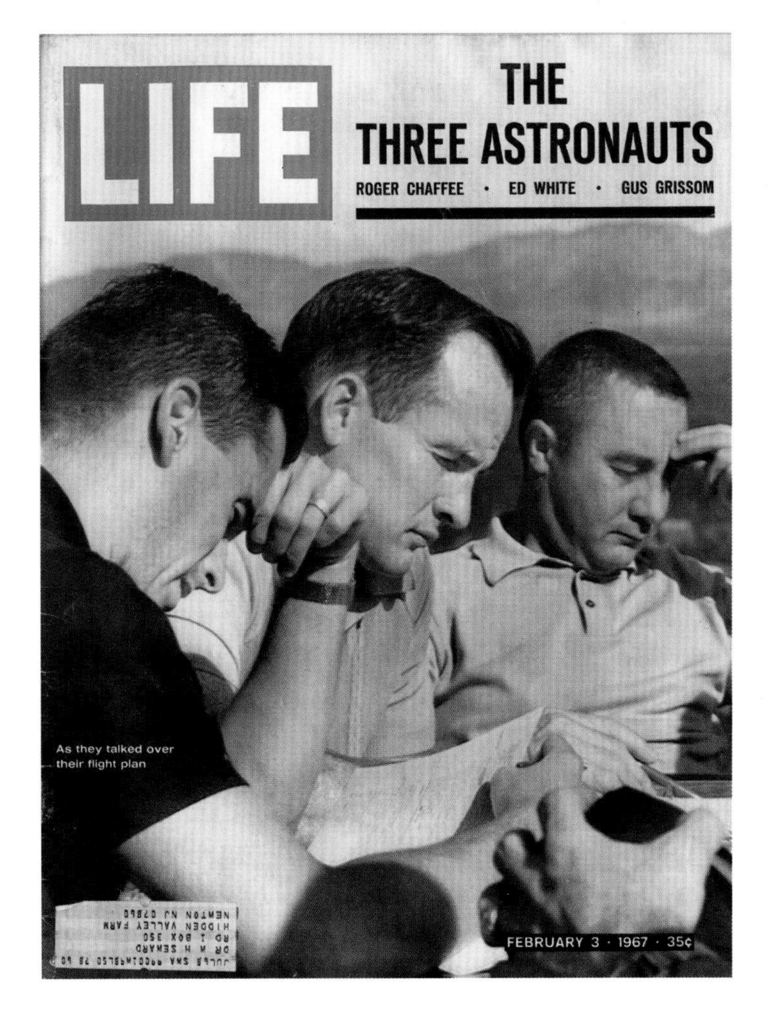

LIFE
THE
THREE ASTRONAUTS
ROGER CHAFFEE · ED WHITE · GUS GRISSOM
As they talked over
their flight plan
FEBRUARY 3 · 1967 · 35¢

The Apollo 1 fire. After the tragic fire on January 27, 1967, that killed Apollo 1 astronauts Gus Grissom, Ed White, and Roger Chaffee, the Apollo program ground to a halt while the cause of the fire was investigated. In the aftermath of the accident, internal NASA/contractor communications and coordination were put under scrutiny through a program known as TIE (Technical Integration and Evaluation), in which NASA relied on one of its chief contractors, Boeing, for implementation and coordination. The Public Affairs Office's media practices, which had been severely criticized, were also reexamined, as were internal NASA-contractor communications.

16. Harold Carr, interview with the authors, December 14, 2011.

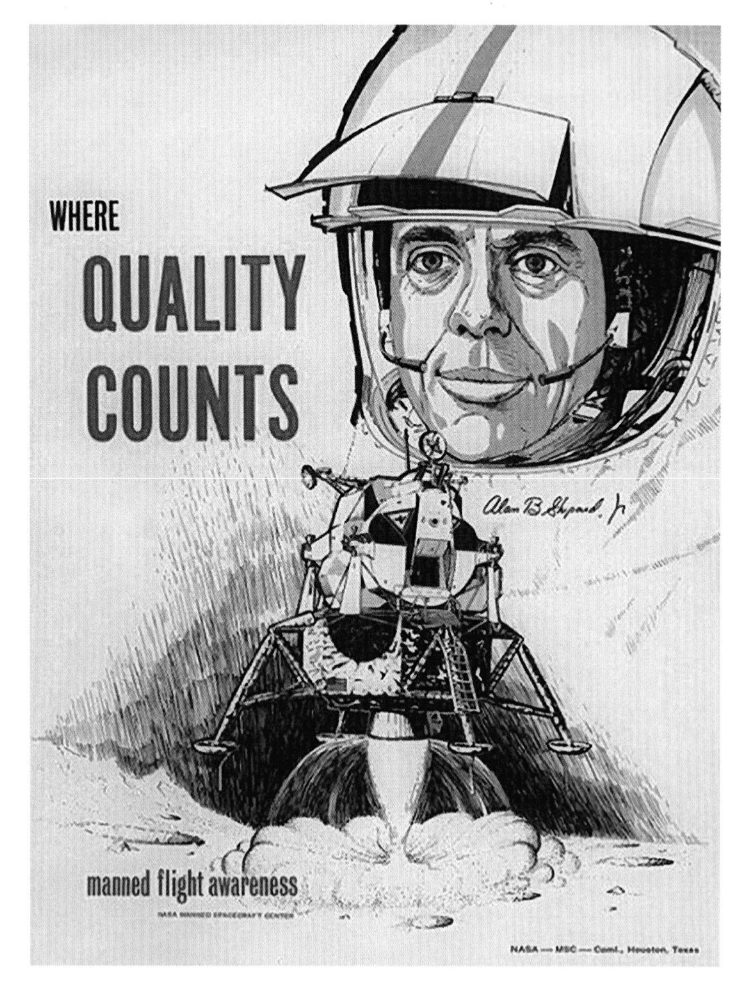

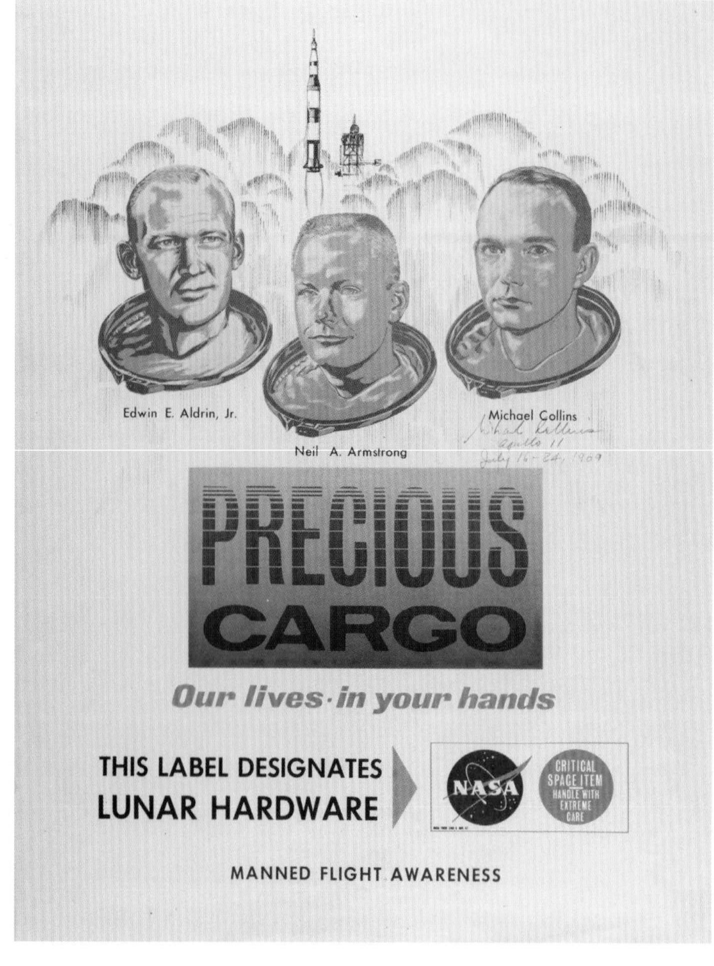

The Manned Flight Awareness (MFA) program received intense attention after the fatal Apollo 1 fire. The program was a collaborative effort of NASA and the many contractors and subcontractors to focus on quality and error-free production of critical parts and equipment. Taking a page from the colorful "loose lips sink ships" World War II posters, MFA posters were used at contractor and NASA facilities to remind employees of the names and faces of the astronauts whose lives depended on the smooth operation of the parts and equipment they were manufacturing.

The problem was well known to members of the news media, who were aware of friction among the NASA Public Affairs people. Mark Bloom described what he witnessed: "There were rivalries among the NASA centers and they all had their PR people. Marshall had their PR person; the Michoud Assembly Facility located at Marshall had their own PR person as well; of course Houston had a battery of PR people; headquarters in Washington D.C. had PR people; Goddard Space Flight Center had PR people; JPL [Jet Propulsion Laboratory, in Pasadena] too. All these people had a stake in Apollo and they were all rivals for credit. Of course, the contractors were also rivals for credit against NASA and among themselves. So everybody was trying for a piece of the action."[17]

TIE concluded that the power structure within NASA was a contributing factor to the disaster, and that good communications were crucial to the success of the Apollo program. To that end, Boeing worked with other contractors, including Xerox, American Telephone and Telegraph, and Western Union, to establish the Blue Teleservice Network, a dedicated system that linked key NASA managers at various facilities via telephone, telegraph, radio, and high-speed data fax.

Boeing and NASA didn't want any publicity about TIE during the investigation. While in most other activities NASA Public Affairs officers and contractor PR people encouraged media contacts to generate positive stories in the press, TIE was the exact opposite: they worked to keep the project out of the spotlight while the team attempted to correct deficiencies within to the program. Yet, Apollo TIE was not without controversy. Many at the Kennedy Space Center criticized it as an unnecessary and expensive public relations scheme to convince Congress of NASA's sincerity at promoting safety, and insisted that the study should have been conducted internally.

17. Mark Bloom, interview with the authors, January 13, 2012.

Some questioned whether it was proper for a NASA contractor to be paid to investigate the work performed by that same contractor, in this case Boeing's construction of the first stage of the Saturn V.[18]

In retrospect, most Apollo-era veterans agree that TIE contributed greatly to success of the program. Boeing's contribution—both coordinating the project and making its talent available—underlines just how completely the space agency treated contractors as a part of the NASA organization. The changes to agency communications implemented as a result of the TIE project led to the revised Apollo spacecraft hardware being in the best possible working order.[19]

Three weeks after the Apollo fire, as the Boeing Apollo-TIE contract was being written, a NASA program manager passionately addressed a gathering of quality control officials and the chairmen representing NASA's major hardware contractors. "It seems to me that within the elements of what happened here lay the seeds of a really significant motivational program for the workers in your plants," he told those in the room.

"Somehow, we have to reach a degree of sophistication in the motivational area that I have yet to see," he said. Even though exhausted after long hours on the investigation into the accident, he attempted to emphatically articulate his message. "It's got to get the in-line people in the organization feeling responsible, to get them to really understand their job, to get them feeling they're a part of the thread the leads right up to the launch."[20]

Behind Apollo, there was the long supply chain of hundreds of thousands of employees at thousands of contractor and subcontractor companies. Making sure everyone involved was motivated, committed, and working in orchestrated concert would require major commitment throughout the system. The scale of the challenge was daunting. For example, the Saturn V alone consisted of more than 3,000,000 parts, ranging from nuts and bolts to circuit boards, washers, and transistors. The command and service module had nearly 2,000,000 parts; the lunar module, 1,000,000.[21] A single mistake in craftsmanship or design, or a less than thorough quality-control inspection, could result in a catastrophic failure, possible loss of life, and a colossal PR nightmare.

"The people that were working on the program were giving us 110 percent of their capability every day," remembered Christopher C. Kraft, Jr., Director of Flight Operations during the early Apollo flights, and later Deputy Director of the Manned Spacecraft Center in Houston. "I don't think we could have done it without that kind of dedication to the program." Kraft acknowledged what became painfully obvious as a result of the Apollo 1 fire: such an effort could not have been achieved or sustained without immense coordinated communication efforts and programs on the part of NASA and the contractors.[22]

Even though each contractor and NASA facility had its own, individual initiatives—from company newsletters and magazines to award and employee-recognition programs—no other program had as much impact, or was as widely embraced across the entire supply chain, as the Manned Flight Awareness Program. Formally launched as a NASA program at the Manned Spacecraft Center in April 1966, the program

Aerospace news coverage published immediately after the 1967 Apollo 1 fire focused on the tragedy's causes and the investigations of the Apollo Accident Review Board. Some journalists wrote pieces highly critical of North American Rockwell, the prime contractor of the command module, though later accounts deemed NASA responsible for most of the design decisions that led to the fire.
A week after the successful first manned Apollo mission, North American Rockwell ran this full-page in Newsweek, *a confident celebration of the nearly flawless mission of Apollo 7 and the redesigned spacecraft. The illustration emphasizes the three large main parachutes, a visual symbol suggesting the importance of safety and protection in a program that had appeared prone to unexpected dangers a year-and-a-half earlier. The advertisement also served to reassure shareholders and the American taxpayers that the company was proud of its role as a prime contractor, and looking forward to playing a central role in "the big trip" to come.*

18. Charles D. Benson and William Barnaby Faherty; *Moonport: A History of Apollo Launch Facilities and Operations.* NASA Special Publication-4204, 1978.
19. Boeing historical information found at company website and in an interview with retired Boeing public relations executive Harold Carr.
20. "MFA stresses quality in every job leading to manned launches," JSC Newsletter, July 11, 1969, p. 7.
21. Roger E. Bilstein, *Stages to Saturn: A Technological History of the Apollo/Saturn Launch Vehicles.* NASA SP-4206, 1980, 1986, p. 288.
22. Christopher C. Kraft, Jr., Johnson Space Center Oral History Project Interview. May 23, 2008, p. 42.

became a joint industry/NASA-wide imperative after the fire. A Manned Flight Awareness office was established at each of the manned spaceflight centers, and by the launch of Apollo 11, most of the largest hardware contractors had established formal Manned Flight Awareness programs, "disseminating program information from the prime-contractor level all the way down to the smallest parts suppliers and vendors.[23]

The unique, agency-wide Apollo-TIE study, the Accident Review Board investigation, and Manned Flight Awareness and other contractor motivational programs were essential in moving beyond and refocusing priorities after the January 27, 1967, tragedy. Yet, meaningful recovery often requires informal, less deliberative solutions as well. Long hours, close friendships, and shared goals had forged a closely integrated community of NASA employees, contractors, and the press during the Mercury and Gemini years. Immediately after the trauma, everyone was painfully aware just how easily and how quickly things could go wrong. Morale at NASA and among its contractors was fragile.

"The Hiatus Was Officially Over"

At this crucial turning point in America's quest to reach the Moon occurred one of the more remarkable events in the Apollo saga. The venue was the Escape Velocity Press Club, Houston's informal, invitation-only gathering spot where journalists, NASA Public Affairs officers, and public relations specialists attempted to escape the "gravity" of their Earth-bound pursuits over a few after-hours drinks. Created as a private club during the early days of Gemini as a way to bypass Texas's restrictive liquor laws, the Escape Velocity Press Club operated out of various hotel meeting rooms whenever the press descended on Houston to cover a manned mission.

In the months after the Apollo 1 fire, NASA flew no manned missions, a dramatic contrast to Gemini, which had flown ten missions in less than two years. After working on stories about a new mission nearly every other month, journalists and public affairs people were suddenly without a flight to talk about. NASA veteran Bob Button recalls the moment: "After the fire, Jim Schefter, the president of our little Escape Velocity Press Club, and I went to see Al Shepard, then head of the Astronaut Office in Houston. (I was then the NASA Public Affairs officer assigned to the Astronaut Office.) Al had been saying, 'It's time folks stopped whining about Apollo 1. Enough is enough.' He told us, 'We've got to get our spirit back and get our asses back in space.'"[24]

Button and Schefter proposed an Escape Velocity Press Club gala, with the upcoming sixth anniversary of Shepard's Mercury flight as a great excuse to party. Shepard was widely known to dislike the press and reluctantly agreed to interviews. Yet, he recognized the value of lending his name to an event that might motivate journalists to get out of their collective funk.

The gala was held in a large ballroom at the Nassau Bay Hotel, located across the street from the Manned Spacecraft Center, in Houston. The room had a capacity of 300, but more than 500, including many VIPs, wanted to attend. Escape Velocity Press Club members had first dibs on the available seats, so Button and Schefter were forced to pick and choose who would attend. "It fell to us to be the bad guys," Button recalled years later. "We turned away some big names to our everlasting agony—entertainers, captains of industry, state and local politicians, friends of astronauts."

Jim Schefter served as emcee, and at the head table were seated Bob Gilruth, director of the Manned Spacecraft Center, and all six of the surviving Mercury astronauts. The widows of Grissom, White, and Chaffee sat at special tables up front. Bob Button served as Pat White's escort, and Paul Haney escorted Betty Grissom.

"While the party was supposed to be an anniversary celebration of Al's flight, it came off more like a roasting of Alan B. Shepard," Button remembers. "And he loved it. We all did! This was the social event of the decade at the Manned Spacecraft Center, and people left that place full of new energy and enthusiasm for getting America back into space and on route to the Moon. The hiatus was officially over." ⊙

23. JSC Newsletter, July 11, 1969, p. 7.
24. Bob Button, interview with the authors, March 13, 2012.

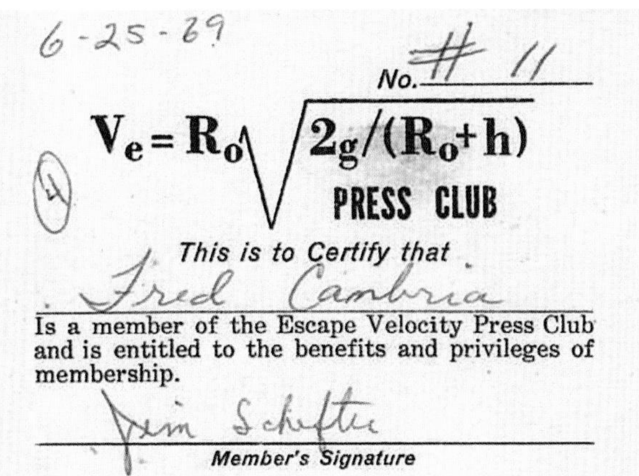

6-25-69

No. # 11

$$V_e = R_o\sqrt{2g/(R_o+h)}$$

PRESS CLUB

This is to Certify that

Fred Cambria

Is a member of the Escape Velocity Press Club and is entitled to the benefits and privileges of membership.

Jim Schefter

Member's Signature

The Escape Velocity Press Club.

Upper left: *Alan Shepard addresses the group (note the bag of Royal Oak charcoal on the table, which was likely a prop in the "roast").*

Lower left: *This Escape Velocity Press Club member's card was issued to Fred Cambria, who worked for CBS television. Jim Schefter, who was president of the club, signed the card, which is dated just a month before the launch of Apollo 11.*

Top right: *An informal singing quartet of astronauts: Dick Gordon, Pete Conrad, John Young, and Tom Stafford.* ***Center:*** *Betty Grissom, widow of astronaut Gus Grissom; to her left is Bob Button; behind her is Wernher von Braun, and to her right, back turned, is Paul Haney.* ***Bottom:*** *Wernher von Braun.*

OFF
MAIN MOM
TALK

The Whole World Is Watching: Live TV on Apollo

Nobody ever said it because nobody had to say it, but I always figured that there was an understanding between television and NASA—never spelled out, never even whispered, never even hinted at, but they knew and we knew. If we continued to help the space agency get its appropriations from Congress, they would in turn give us, free of charge, the most spectacular television shows anyone had ever seen. Those of us who produced the television coverage of space soon found out that what we were getting was a lot more than free television shows. What we were getting as well was a chance to show the American people that we were team players, and that if television brought you distasteful things like race riots and a war we couldn't win, it also brought you the astronauts. It was a chance for television to show that it, too, had the right stuff. —Don Hewitt, CBS News [1]

AMERICANS TUNING IN to any of the three network newscasts the evening of Monday, October 14, 1968, were confronted with stories and images representative of the tenor of that divisive and tumultuous year. The heated presidential contest between Richard Nixon and Hubert Humphrey was entering its final weeks. In the aftermath of the disastrous and violent Democratic convention in Chicago seven weeks earlier, political analysts were expressing concern about the fate of the sitting president's party, should George Wallace's populist, third-party bid force the election into the House of Representatives. The extended length of the war in Vietnam, the catalyst for much of the national unrest, prompted the Department of Defense to announce that 24,000 soldiers and Marines would be assigned to involuntary second tours of duty. In Mexico City, where the XIX Olympiad had opened during the previous weekend, reporters were sensing increased political tension on the athletic fields, prompted by the Soviet invasion of Czechoslovakia in August and growing racial unrest in the United States. On October 16th, the Olympiad was the site an iconic moment in sports history, when American sprinters Tommie Smith and John Carlos, members of the Olympic Project for Human Rights, defiantly lowered their heads and raised black-gloved fists in a "black power salute" during the playing of the Star Spangled Banner.

That evening, the network newscasts also presented a sight never before seen on American screens: excerpts from the first live television transmission from an American spacecraft orbiting the Earth. Apollo 7, the first manned Apollo flight, had been launched the previous Friday. Fuzzy black-and-white images showed astronauts Wally Schirra, Donn Eisele, and Walt Cunningham moving weightlessly within the cramped cabin of the command module as they conducted a brief tour and held up cards with prepared messages like "From the Lovely Apollo Room high atop everything" (a comic allusion to a much-used phrase that opened popular music radio broadcasts during the 1940s). These unprecedented glimpses of American astronauts relaxed, unscripted, and seemingly at home (and having fun) while at work, redefined the personality of the astronaut corps, and instantaneously dispelled any lingering perceptions of the astronauts as colorless and aloof.

Indeed, the first seven-minute broadcast of what was dubbed by the press "The Wally, Walt and Donn Show" easily slipped right into the daytime entertainment programming of the networks. CBS interrupted its morning network rerun of *The Beverly Hillbillies*, while NBC cut away from *Concentration*, a quiz show hosted by Hugh Downs.

The October 14th broadcast was equally revelatory as an indication of two ongoing revolutions within the telecommu-

OPPOSITE: Astronaut Alan Bean, lunar module pilot for the Apollo 12 mission, holds a Special Environmental Sample Container filled with lunar soil collected during the extravehicular activity (EVA), in which astronauts Pete Conrad and Bean participated. Conrad, the commander, who took this picture, is reflected in Bean's helmet visor.

1. Don Hewitt, *Tell Me a Story: Fifty Years and 60 Minutes in Television*. New York: Public Affairs, 2001, p. 73.

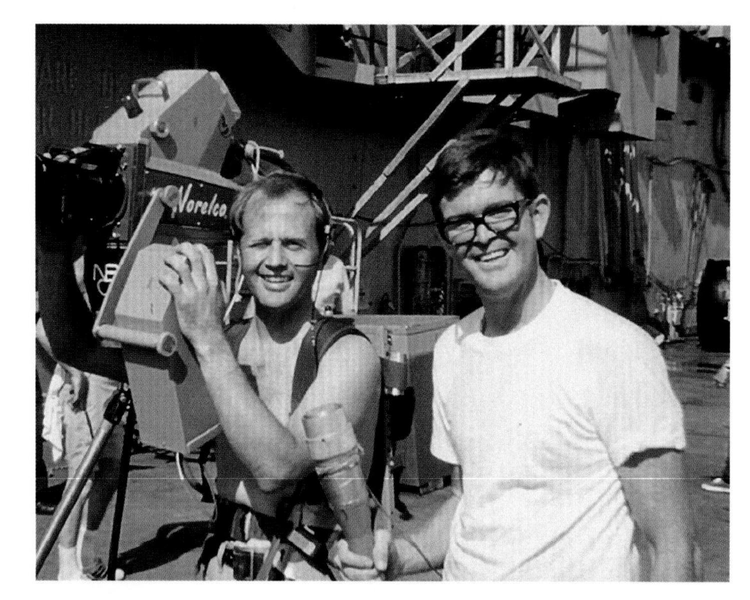

Upper: Two NBC News employees with state-of-the-art live television equipment of 1969. The cameraman on the left is supporting the Norelco PCP 70 camera, a commercial model of the CBS Minicam Mark VI, the first hand-held color TV camera system. The PCP 70 weighed more than 50 pounds and was nicknamed "the backbreaker" by those who operated it. The NBC crew shown here are aboard the USS Ticonderoga for the May 1969 splashdown of Apollo 10.

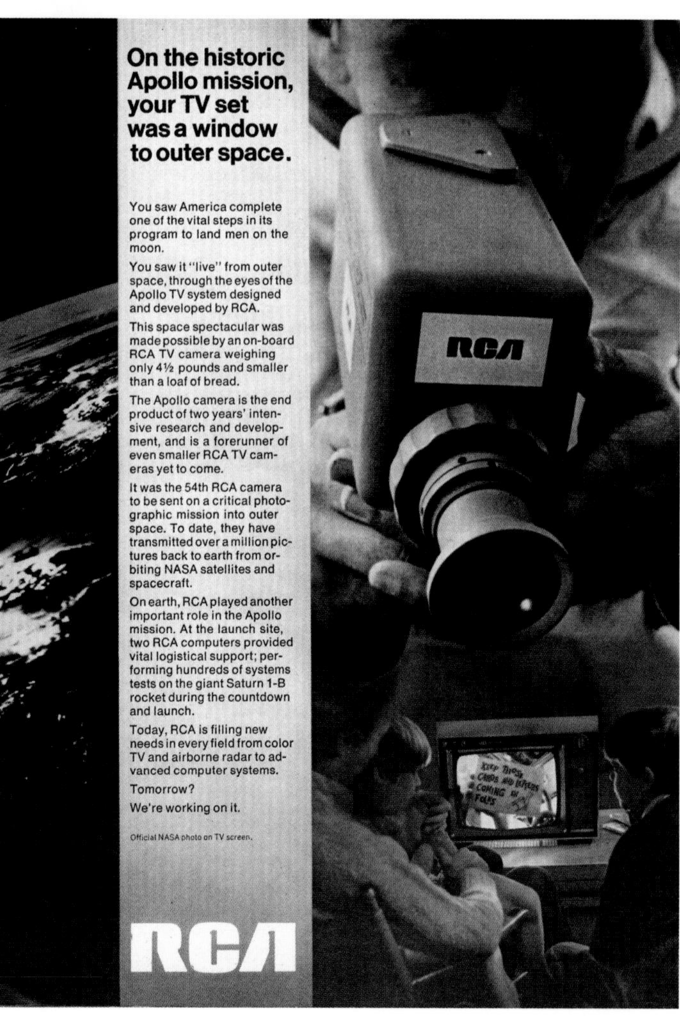

On the historic Apollo mission, your TV set was a window to outer space.
You saw America complete one of the vital steps in its program to land men on the moon.
You saw it "live" from outer space, through the eyes of the Apollo TV system designed and developed by RCA.
This space spectacular was made possible by an on-board RCA TV camera weighing only 4½ pounds and smaller than a loaf of bread.
The Apollo camera is the end product of two years' intensive research and development, and is a forerunner of even smaller RCA TV cameras yet to come.
It was the 54th RCA camera to be sent on a critical photographic mission into outer space. To date, they have transmitted over a million pictures back to earth from orbiting NASA satellites and spacecraft.
On earth, RCA played another important role in the Apollo mission. At the launch site, two RCA computers provided vital logistical support; performing hundreds of systems tests on the giant Saturn 1-B rocket during the countdown and launch.
Today, RCA is filling new needs in every field from color TV and airborne radar to advanced computer systems.
Tomorrow?
We're working on it.
Official NASA photo on TV screen.
RCA
RCA

Lower: RCA published this consumer advertisement in national magazines shortly after the conclusion of the Apollo 7 mission. At a moment when RCA was trying to sell new color TV sets to millions of consumers, the implied message was clear: purchase your next home television from the company chosen by NASA. To 21st-century eyes, the camera looks like a typical artifact from the early era of home video, in the 1980s, but in the late 1960s, when the advertisement was created, such devices were still far off in the future for home use.

nications industry during the 1960s. The first revolution was technological, as new innovations—light, portable 16mm news cameras with synced sound; international communications satellites; refined color television; and solid-state electronics—gave news gatherers opportunities to capture images and sound with immediacy, clarity, and ease. The second took place on the airways, as these technological breakthroughs dramatically changed the very nature of television news. Seven years earlier, in 1961, when man first orbited the globe, a fifteen-minute network evening newscast was little more than a summary of wire service dispatches illustrated with still images and occasional newsreel footage (usually without synced sound). By 1968, the three major networks had expanded their newscasts to a half-hour format. Following the launch of the Telstar, Syncom, and Early Bird satellites, live international television broadcasts were no longer unusual. (Millions of Americans witnessed the state funeral of Winston Churchill, broadcast live in January 1965.) However, producing live television news spectaculars still required the use of cumbersome, heavy cameras, countless electrical cables, and mobile facilities with generators. Network news reports from Vietnam were shot with new, relatively lightweight 16mm cameras; the film and accompanying audio was dispatched by air from Saigon to New York City, where the film was developed, synced, edited, and rushed to broadcast a day or two after it was shot. Hand-held live color television cameras were still a novelty. CBS News made history with the Mark VI at the 1968 Democratic and Republican conventions, allowing viewers to watch over the shoulders of their correspondents as they conducted exclusive live interviews with delegates and other VIPs.

Even within this rapidly evolving media environment, live television from an orbiting spacecraft—albeit limited to black-and-white—was an astounding technical achievement. In ads published shortly after Apollo 7's return, RCA, the camera's contractor described the on-board camera as "weighting only 4½ pounds and smaller than a loaf of bread … the end product of two years' intensive research and development and a forerunner of even smaller cameras yet to come."

Ironically, the Apollo 7 television spectacular, and all the live television transmissions from subsequent Apollo missions to the Moon nearly didn't happen. Live television transmissions were not in the original plans for Apollo. Many administrators at NASA said television was unnecessary. Engineers argued that live video was a waste of valuable resources. And most of the original astronauts and their bosses insisted that operating television cameras would detract from the important work of the mission.

Television made it on board the Apollo missions only as a result of the efforts of a small group of NASA visionaries, dedicated teams of engineers, and years of persistence.

Amazingly, during the very first manned spaceflight, a video camera went along for the ride. No live television images were broadcast to the rest of the world when Yuri Gagarin's *Vostok 1* orbited the Earth on April 12, 1961; in front of him, however, was a Vidicon-tube camera shooting pictures at ten frames per second (100 lines per frame). The Soviets, experienced at capturing the world press's attention with surprise space firsts since *Sputnik 1*'s launch in 1957, offered up recorded images from the Vostok's video camera to reinforce their achievement, and to make palpable the U.S.S.R's technological prowess. One of the primal conflicts waged during the Cold War was the battle of images, waged in the world press. Wire photos and newsreels showed the world the latest evidence of military power, material plentitude, or technological genius—images and film clips that vied to prove the superiority of Capitalism or Communism and sway opinions in the developing world. To top their space triumph, the Soviets found a way to garner even more attention. That day the Soviet Union allowed the world to see the first-ever live broadcast from inside Russia: Premier Nikita Khrushchev welcoming the returning cosmonaut with a bear hug and a visit to Lenin's Tomb in Red Square.

As part of the concluding flight of Project Mercury in May 1963, NASA conducted its first experiment with a slow-scan television camera on an American manned spacecraft. However, neither the pilot, Gordon Cooper, Jr., nor his experience on *Faith 7* did much to encourage further television development. The semi-portable camera, designed and built by Lear Siegler Inc., was contained in a Thermos-like housing attached to a cable and bracket. The equipment added an additional 17 pounds to the weight of spacecraft and operating it consumed a demanding 56 watts.

While in orbit, Cooper conducted more than twenty tests of the camera, and though some images were received showing Cooper while weightless, the video system failed to work many times. After the flight, Cooper noted that the camera was ungainly, difficult to operate, and interfered with his ability to conduct other onboard activities. In a remark that only served to strengthen the anti-television faction within NASA during the years that followed, Cooper noted that he "could see no real advantages to the pilot in having it onboard."[2]

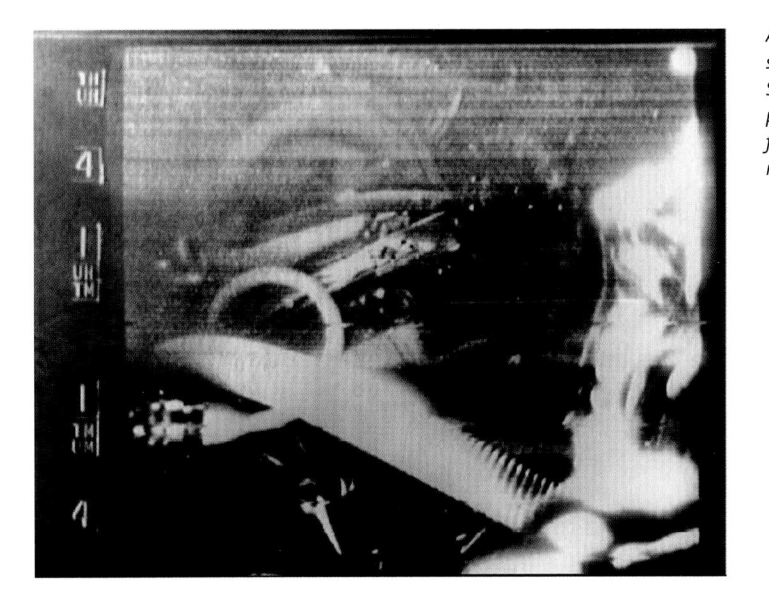

Astronaut Gordon Cooper as seen in a slow-scan video aboard Faith 7. The Lear Siegler camera transferred ten frames per second with a resolution of 320 frame lines using less bandwidth than a normal TV transmission

Cooper's fellow Mercury astronauts Wally Schirra and Deke Slayton were among those most opposed to live television in a spacecraft. "The live TV idea was propaganda, pure and simple," Schirra recalled in his memoir. "The astronauts remained neutral, but we'd resist anything that interfered with our main mission objectives."[3] Slayton, grounded by a medical condition in 1962, assumed a new and influential position as NASA's Director of Flight Crew Operations, and in this role he advocated against adding seemingly superfluous tasks to the flight plan.

On a second front, the engineers at NASA and at the assigned contractors resisted the inclusion of television on Apollo, since a camera and the associated signal delivery hardware would only add to the weight to the spacecraft, particularly the lunar module. (Indeed, weight became so critical an issue during the lunar module manufacture that NASA paid Grumman, the prime contractor, $50,000 for each pound it removed from total launch weight.) Tasked with designing a reliable vehicle to safely transport men to the Moon and back,

2. Dwight Steven-Boniecki, *Live TV From the Moon*. Burlington, Ontario: Apogee Books, 2010.
3. Wally Schirra and Richard N. Billings, *Schirra's Space*. Boston: Quinlan Press, 1988, p. 202.

the engineering community considered television an unnecessary extra, unrelated to the functional operation or objectives of the mission.

Quietly, NASA encouraged a number of outside contractors to submit proposals for a compact, portable television system capable of transmitting live broadcasts from within the Apollo spacecraft and from the lunar surface. A NASA report dated April 3, 1962, disclosed, "of ten proposals received . . . no contractor met the full requirements of the specifications."[4]

Little more than a year later, word appeared in a North American Aviation progress report that RCA Princeton Laboratories had been chosen as the subcontractor responsible for the production of the Apollo command module television camera. The report noted that, not only would the planned RCA camera broadcast images from the command module, but it would be modified to operate compatibly with the systems being developed by Grumman for the lunar module and, therefore, offer the possibility of providing live images from the lunar surface as well. Yet the RCA camera referred to in this 1963 report never made it to the surface of the Moon. That honor went to a camera developed by Westinghouse, which was awarded a $2.29 million contract in 1964 to create a camera specifically designed to operate in the lunar environment.

Thus, during the early 1960s, one might encounter an occasional press reference to the possibility of a live television transmission showing man's first steps on the Moon, but there was very little public discussion to encourage such speculation. Certainly, if Project Gemini was any indication, television would be absent from the Apollo flight plan. Gemini's series of two-man missions during 1965 and 1966 were designed for future Apollo astronauts to test equipment, master tasks, and perform maneuvers necessary for a lunar voyage, such as executing a successful rendezvous and docking with another vehicle, enduring flights of extended duration, and working independently outside the spacecraft. The only cameras on board during Gemini were conventional still frame and 16mm movie cameras.

Nevertheless, within NASA a handful of visionaries remained convinced of the visual power of live television. Major proponents within NASA included George Low, manager of the Apollo Spacecraft Program office, and Christopher Kraft, director of Flight Operations. Playing an equally crucial role was NASA Public Affairs chief Julian Scheer and his team, particularly Jack King and Robert J. Shafer.

In the opinion of political scientist John M. Logsdon, author of *The Decision to Go to the Moon: Apollo and the National Interest,* Project Apollo was as much about "projecting a positive image of the United States as it was about science and exploration . . . [and] Scheer was key to the success of that effort."[5]

Scheer took his case directly to the engineers. "Weight was a critical issue, no question about it," he said. "But I insisted, 'You're going to have to take something else off. That camera is going to be on that spacecraft.'" In response, the engineers fought back. "'No, no, you don't understand. It'll interfere with flight qualifications.'" Scheer recalled them insisting, "'Our job is to get the astronauts to the Moon and back safely, and bring a [soil] sample back—not to appear on television.'"[6]

Among the astronauts, Tom Stafford, a member of the second group and a veteran of two Gemini flights, voiced his disagreement with Gordon Cooper and many of the original Mercury 7. "One thing that always surprised me about Deke Slayton and most of the Mercury astronauts was their indifference—or animosity—toward the public affairs side of the manned space program," Stafford wrote some years later.

In his position as head of mission planning for the Apollo astronaut group, Stafford became a major advocate for live television. "It was clear to me that the American public was paying for Apollo and deserved as much access as it could get. They should see the wonders we saw. Hasselblad photos and 8mm [*sic*] movies were great, but nobody saw them until after a mission was over. What better way to take viewers along to the Moon than by using color television?"[7]

Kraft and Stafford, together with Scheer and the others on the Public Affairs team, fought years of resistance. "I even got a note from Deke Slayton that said, 'We're not performers, we're flyers,'" said Scheer. "They could never see the big picture. But they weren't landing on the Moon without that [TV] camera on board. I was going to make sure of that. One thing I kept emphasizing was, 'We're not the Soviets. Let's do this thing the American way.'"[8]

During the mid-20th century, no artist was better known for depicting the American way—or, perhaps, America's idealized vision of its way—than Norman Rockwell. So to subtly

4. Dwight Steven-Boniecki, *Live TV From the Moon*, Apogee Books, 2010, p. 24.
5. Quotation in obituary for Scheer, *The New York Times*, September 5, 2001.
6. Billy Watkins, *Apollo Moon Missions: The Unsung Heroes*. Lincoln, Neb.: University of Nebraska Press, 2007, p. 51.
7. Thomas P. Stafford and Michael Cassutt, *We Have Capture: Tom Stafford and the Space Race*. Washington, D.C.: Smithsonian, 2002.
8. Watkins, op. cit., p. 52.

Norman Rockwell's painting "Man's First Step on the Moon," commissioned by NASA, was first published in Look *magazine, January 10, 1967. It was based on photographs taken by Molly Rockwell of the scene set up by her husband (the thin, gray-haired man) near the Manned Spacecraft Center, in Houston, where a full-scale model of the lunar module was put at the artist's disposal for the session. Upon publication, the painting was criticized for a number of technical inconsistenices: the size of the Earth is too large in comparison to the Big Dipper, the command module and the Earth are illuminated from different directions, and the lighting on the lunar surface is further inconsistent. Rockwell was obsessed with accuracy and, for this project, he collaborated with space artist Pierre Mion, seen in the lower left of the six photos. It may have been Mion who urged Rockwell to think dramatically rather than worry about certain details. What is perhaps most remarkable about the painting is the accuracy with which it foretold Neil Armstrong's actual descent onto the lunar surface more than two years later.*

Note that, both in the painting and in the photograph at the lower right, the "astronaut" is holding a mock-up of the Westinghouse B&W camera. This was very likely the result of someone in NASA Public Affairs advising Rockwell to include the camera in his painting.

Between the time the photos were taken and the final painting was executed, the round hatch design of Grumman's lunar module was replaced by a rectangular one. Rockwell was furnished with a model of the revised design and that is what appears in the painting.

Reproduced by permission of the Norman Rockwell Museum Collections.

sell their idea to the public, the pro-television faction got some assistance from America's favorite painter. In 1966, NASA commissioned Rockwell to depict the moment of man's first footstep on the Moon. For research, Rockwell traveled to the Manned Spacecraft Center, where he observed and photographed a full-size mock-up of the lunar module and posed models outfitted in the latest Apollo lunar spacesuits.

Rockwell's dramatic painting was first published, with considerable fanfare, as a double-page spread in the January 10, 1967, issue of *Look* magazine. At the time, Americans were most likely to encounter the new issue of *Look* in their dentist's waiting room or at a hair salon. Across the country, pages from the magazine with images from Project Apollo were often used to adorn school bulletin boards. Notable in the upper portion of Rockwell's painting is the lunar module pilot aiming a television camera. Since the painting was a NASA commission, and Rockwell's reference photographs depict a suited Apollo astronaut posing with a prototype of the Westinghouse lunar camera, the inclusion of the camera in the composition was clearly intentional—very likely stage-managed—by someone at NASA. With the mass reproduction of this painting, the pro-television faction cleverly marketed to millions of Americans a dream that they, too, would be a witnesses to the monumental event pending in a few months.

"The Final Impossibility: Man's Tracks on the Moon," painting by Norman Rockwell, 1969, This painting was released following the first Moon walk, on July, 20, 1969. Here, the camera is no longer a detail, but an essential element of the scene. (See Ron Schick, Norman Rockwell: The Man Behind the Camera. *New York: Little, Brown, 2009.) Reproduced by permission of the Norman Rockwell Museum Collections.*

Of the vocal proponents within NASA, Christopher Kraft held one of the highest positions. "I can't conceive of this country 'sending' three men to the Moon and not being allowed to see the lunar surface and the sight of a U.S. LEM on the Moon," Kraft wrote in a memo to Slayton. "I believe you should reconsider your point of view."

NASA was also receiving external pressure from network producers, who were constantly looking for new ways to make their coverage more visually alluring. After NASA had allowed the first live television images from a Gemini recovery ship, the networks requested a camera in a recovery helicopter. Again Slayton objected, telling Public Affairs officer Brian Duff, "I never want anyone to see an American astronaut losing his lunch on that spacecraft." Eventually it took the will of Manned Spacecraft Center Director Robert Gilruth to make Slayton back down, and once overruled, Slayton never argued the question of television again.[9]

The final obstacle disappeared shortly before the flight of Apollo 7, when Julian Scheer overcame objections against live television voiced by Wally Schirra and George Low at a conference at Ramey Air Force Base in Puerto Rico. "The Ramey decision" as Scheer referred to it "was a real watershed for the space program."[10]

Ironically, just as live television from Apollo had proved itself and brought the population of the planet together during that unforgettable moment in July 1969, the technology appeared to lose its magic. The reasons why have been debated for decades. The live television camera mishap on Apollo 12 was a key factor in the decline. As soon as the pictures of Pete Conrad and Alan Bean conducting their EVA at the Ocean of Storms went dead, the public lost interest and tuned out, a clear indicator of live television's power to generate interest and motivate enthusiasm. The reduced number of viewers during the compromised Apollo 12 coverage hardly went unnoticed at the networks, giving them the rationalization to gradually reduce their commitment.

Additionally, the goal of landing a man on the Moon was often thought of as a Cold War competition with a clearly defined finish line. Thus for many, the missions following Apollo 11 appeared anticlimactic, redundant, and costly. Apollo 12 through 17 also had the unfortunate fate of occurring at a moment in American history when the nation was redefining itself following the tumultuous social and cultural shifts of the late 1960s, and during the excruciating final years of the Vietnam War. Simultaneously, there was a growing concern about the fragility of our home planet, sparked in part by the profound effect of the Apollo 8 "Earthrise" photograph taken in December 1968. (The first Earth Day took place less than eighteen months later, on April 22, 1970.)

Even greatly refined television images from the Moon during the final three Apollo missions, which included lunar rovers, did little to recapture the enthusiasm of the general public during the early 1970s. As the result of massive cutbacks by the three commercial networks, most Americans saw very little of the television broadcast from the Moon during the final two Apollo missions.

From a vantage point four decades after the final Moon landing, the failure of the television networks and the American public to fully appreciate the later Apollo missions is difficult to understand. Only three years earlier, the Apollo program had changed both the world and television forever. Humans never thought of their home the same way after see-

9. Brian Duff quoted in a 1987 interview in Robert J. Donovan and Ray Scherer, *Unsilent Revolution: Television News and American Public Life, 1948–1991*. Cambridge University Press, 1992, pp. 54–55.
10. Julian Scheer quoted in a 1987 interview, ibid.

ing the live transmissions from Apollo or after gazing at the photos of the Earth surrounded by the dark void of space. And the remarkable television and photographic images of human beings walking and working on an alien world forever redefined who we are as a species.

When reflecting upon the importance of television in Apollo's legacy, Gene Cernan, the last man to walk on the Moon, was emphatic. "If you want to market the Apollo program, put the astronauts on television and let them have a press conference on the way home. Without television, Apollo would have been just a mark in a history book. But to those people who were alive and remember it, it's visionary, as if they were there.

"One of the best compliments I got was: 'Gene, you took me with you.' And that's what I wanted to do. I was there with you. That's what television did—it took you with us when we went to the Moon. We didn't say 'We'll tell you about it in two weeks.' We took you with us. The power of television is unbelievable. That's what television does. What you are seeing is happening at this instant. The liftoff from the Moon is a good example. You could sit in your living room listening to the commentary of Apollo 17, but you could also watch it happening at that instant.

"I think if you talk about marketing on a grandiose basis, when you focus on Apollo, the thing that meant so much and brought so much prestige to this country is that every launch, every landing on the Moon, and every walk on the Moon was given freely to the world in real time. We didn't doctor up the movie, didn't edit anything out; what was said was said. From a marketing point of view, I don't think there's anything better."[11]

Astronauts Wally Schirra Jr. (right) and Donn Eisele (left) of Apollo 7 are seen in this first live television transmission from space. Schirra is holding a sign that reads, "Keep those cards and letters coming in folks!" This transmission was made at the end of their third day in space.

APOLLO TV BY MISSION

Apollo 7

The first manned Apollo mission was also the first to broadcast live television from space, despite its commander, Wally Schirra, having been one of the most reluctant to carry a camera on board. The mission involved testing the newly redesigned Apollo spacecraft on its maiden voyage, and there were many tasks to complete in an ambitious timeline. The controllers on the ground and the television networks were eager to have television as quickly as possible and planned the first test for the second day.

Schirra and his crew were focused on their primary assignment and as mission commander, Schirra was particularly attentive about doing things in the precise order stipulated on the checklists. "We're not going on television today," Schirra said to the ground. Deke Slayton came onto the communications link to talk directly to Schirra, to persuade him to change his mind, a rare move for NASA's Director of Flight Crew Operations, the astronauts' boss. But the commander of the spacecraft had the final say: "Sorry, Deke. No TV today."

Part of the reason for Schirra's stubborn refusal may have stemmed from a lack of training in a skill far removed from an astronaut's normal duties. While NASA did a fine job giving the astronauts technical training about the basic operation of the television system, its switches and cable connections, making them comfortable with the creative side of the process

11. Gene Cernan, interview with authors, February 24, 2012.

The RCA slow-scan command module TV camera was used on Apollo 7 and Apollo 8, not only to broadcast images to an eager public, but also to test the use of real-time television feeds from spacecraft in Earth orbit and lunar orbit. The system was designed to send signals over all unified S-band NASA tracking stations while in Earth orbit, and through the three stations with 85-foot antennas located in Canberra, Australia, Goldstone, California, and Madrid, Spain, while broadcasting from lunar distances. As Apollo 8 was the first spacecraft to leave Earth orbit with humans aboard, live television was a critical component of the public relations aspects of the mission and a crucial test of the technologies that would eventually be used with upgraded TV cameras broadcasting images of humans on the surface of the Moon. The combined weight of the RCA camera and its cables and lenses was only 6.2 pounds, remarkably light given the technology of the time. The camera system was developed by RCA for a total contract cost to NASA of approximately $4.2 million. The top two images are of a wooden prototype used by Apollo 8 astronauts in training for their mission; the bottom image is the actual camera.

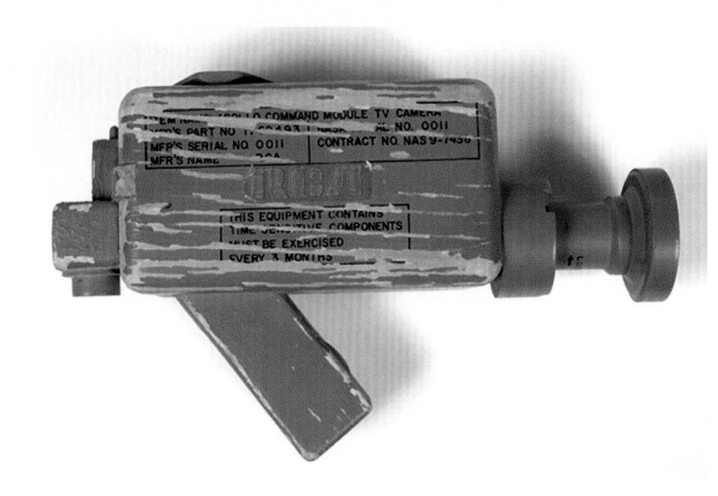

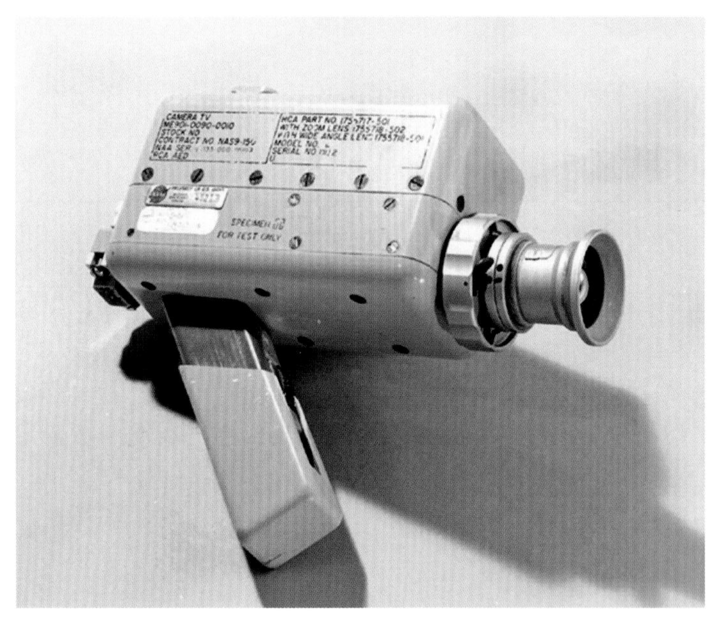

and suggesting how to conduct a successful television broadcast had been ignored completely by the staff in Public Affairs. With no fundamental understanding of the power of television, the crew relegated it to a less important status. "I don't think there was ever a word of encouragement at all," said Walt Cunningham, Apollo 7's lunar module pilot.[12]

But once the "Wally, Walt and Donn Show" debuted on October 14, the crew learned to enjoy the broadcasts. Fortuitously, a series of humorous cue cards had been prepared in advance for the crew to hold up to the camera so they could be read by the capcom in Houston. They also pioneered the "spacecraft tour," replete with images of objects floating in the cabin. Audiences were fascinated to get a glimpse of the first live pictures of the Earth, especially when recognizable landmarks such as the Florida panhandle passed beneath. The broadcasts generated tremendous interest and a year later, the Academy of Television Arts and Sciences presented the crews of Apollo 7, 8, 9, and 10 with a Special Trustee Emmy Award.

The device they used was the direct descendant of the RCA camera subcontracted by North American Aviation during Apollo's nascent years. (By 1964, a group at NASA had taken over direct supervision of all the television camera contracts from North American.) This slow-scan, black-and-white system shot only 10-frames per second with 320 lines of resolution. In comparison, had the American NTSC television standard of 1968 been chosen—525 interlaced lines, 30 frames per second—ten times more bandwidth would have been required.

Futuristic by 1968 standards, the ease of the RCA camera's point-and-shoot design and pistol-grip control foreshadowed the appearance of the first consumer home video cameras of the 1980s. This lightweight camera was fitted with either a wide-angle lens (for shooting the astronauts inside the command module) or a telephoto lens (for shooting the Earth or the Moon).

On screen, Schirra appeared to coolly tolerate the additional eye on board, though his honest feelings about the entire experiment were recorded in his log, dictated during the mission. "I believe that the television should be left as the last, low-priority test objective in relation to any other event that might occur simultaneously. Typically, with a television camera on board, the crew reacted to it, and we fortunately had no

12. Walt Cunningham, interview with the authors, February 14, 2012.

problems occur, but we were paying way too much attention to the TV camera and not the spacecraft. . . . A candid-camera syndrome is a very awkward one to have in a spacecraft."

Less than a year later, Schirra was on camera once again, this time in the role of Special Analyst, seated next to Walter Cronkite during CBS News's live network coverage of the mission of Apollo 11. Jovial, reflective, and sharing unique personal insight into the men and the mission, Schirra contributed greatly to the success of the CBS broadcasts. Just moments after Neil Armstrong set foot on the Moon, the astronaut who fought against the inclusion of live television on Apollo said to both Cronkite and the viewing audience, "Oh, thank you television for letting us watch this one!"[13]

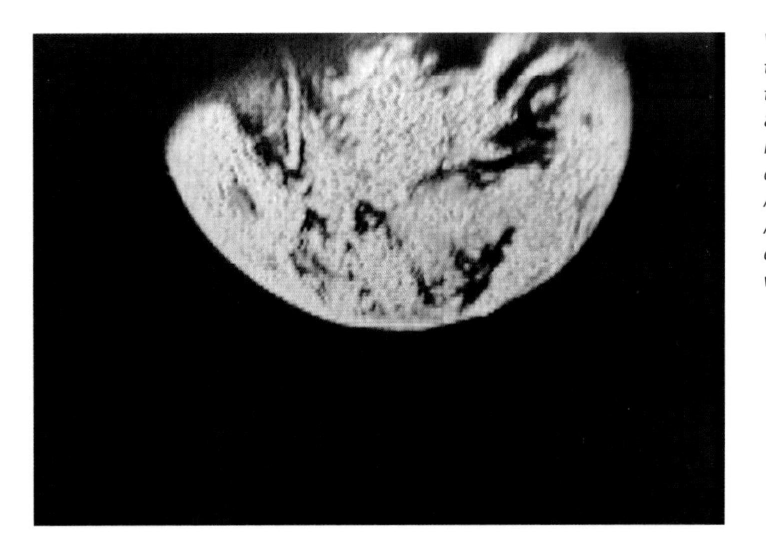

View of the Earth that was transmitted back from space during the live television transmission from the Apollo 8 spacecraft on the third day of its journey toward the Moon. This view was captured through a spacecraft window. At the time of this TV transmission, Apollo 8 was traveling on its translunar course at about 3,254 ft per second, and was 176,533 miles from Earth.

Apollo 8

During Christmas week of 1968, the crew of Apollo 8 became the first humans to venture beyond Earth orbit and travel to the Moon. The engineers who had built the NASA Deep Space Network would have an opportunity to see how it handled the television broadcasts from 250,000 miles away.

Apollo 8's stunning pictures of our tiny and fragile planet seen from the vantage point of the barren and largely colorless Moon affected everyone on Earth. These iconic photographs are now known to generations from the still camera images developed after Apollo 8's return. However, the world's first encounter with the view seen from the window of Apollo 8 came via live television broadcasts utilizing the same black-and-white RCA camera pioneered on Apollo 7. (Notably, Apollo 8 was the last flight that employed the RCA black-and-white camera.) The live television images of a receding Earth were somewhat indistinct. Cities and national borders were invisible. Even distinguishing specific landmasses from among the clouds shrouding the globe took a bit of geographic skill. But the emotional and philosophical impact on viewers was immediate. Viewers sitting at home realized they were looking at a live portrait in which they appeared as little more than a half-illuminated sphere hanging in a black void. And as they watched and reflected upon this sight, the very idea of what it was to be a human inhabitant on the planet Earth changed. Simultaneous with this altered perspective was an additional realization that this was the viewpoint of three fellow human beings, who at that very moment were traveling farther from Earth than anyone since the rise of humankind. For most people living at that moment, the visual realization of the Earth's vulnerability, the stark isolation of the Apollo 8 crew, and the enigmatic and chilly lifelessness exhibited by of our celestial neighbor evoked powerful personal and spiritual emotions. Many writers and thinkers have since suggested that the photographs of the Earth seen from lunar orbit are the greatest legacy of the entire Apollo program.

The culmination of Apollo 8's television spectacular came with their fourth broadcast, on Christmas Eve, December 24, 1968. In what was believed to have been the most watched television broadcast in history up to that moment, the crew of Apollo 8 began their transmission with a wide-angle lens image of the Earth, appearing very tiny and very far away. Then the camera next panned to show the Moon's pitted surface, captured in close detail as the spacecraft passed above in lunar orbit. After describing the appearance of the Moon's surface and some of its features, crew members Bill Anders, Jim Lovell, and Frank Borman then took turns reciting the first ten verses of the book of Genesis from the King James Bible.

BILL ANDERS: We are now approaching lunar sunrise and, for all the people back on Earth, the crew of Apollo 8 has a message that we would like to send to you.

In the beginning God created the heaven and the earth.

13. CBS News live broadcast, Sunday, July 20, 1969.

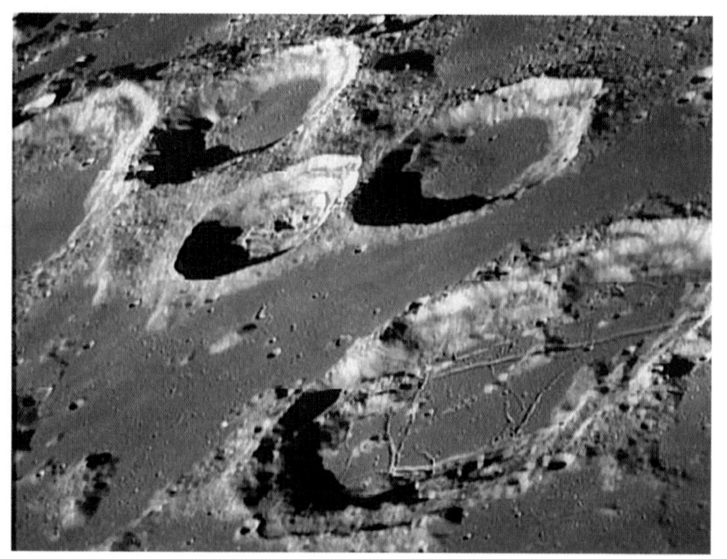

LEFT: View of the Earth as photographed by the Apollo 8 astronauts during their lunar orbital mission. North is about five percent to the right of vertical center. The sunset terminator crosses North America and South America. Clouds cover most of the United States. Only the desert in the southwest and the Florida peninsula are clear.

RIGHT: This photograph was taken from the Apollo 8 spacecraft with a long focal length lens, looking south at the large crater Goclenius, which is in foreground. The three clustered craters are Magelhaens, Magelhaens A, and Colombo A. The crater Goclenius is located at 10 degrees south latitude, 45 degrees east longitude, and it is approximately 40 statute miles in diameter.

And the earth was without form, and void; and darkness was upon the face of the deep. And the Spirit of God moved upon the face of the waters. And God said, Let there be light: and there was light. And God saw the light, that it was good: and God divided the light from the darkness.

JIM LOVELL: And God called the light Day, and the darkness he called Night. And the evening and the morning were the first day. And God said, Let there be a firmament in the midst of the waters, and let it divide the waters from the waters. And God made the firmament, and divided the waters which were under the firmament from the waters which were above the firmament: and it was so. And God called the firmament Heaven. And the evening and the morning were the second day.

FRANK BORMAN: And God said, Let the waters under the heavens be gathered together unto one place, and let the dry land appear: and it was so. And God called the dry land Earth; and the gathering together of the waters called he Seas: and God saw that it was good.

And from the crew of Apollo 8, we close with good night, good luck, a Merry Christmas—and God bless all of you, all of you on the good Earth.

Even as the unique images of the Earth and the Moon provoked awe and wonder, the reading from book of Genesis on Christmas Eve by the first humans to orbit the Moon was a profound moment in the lives of all who witnessed it. Coming at a particularly fragile and frightening moment in world history, the achievement of Apollo 8 united humanity in a singular moment, a moment that also acknowledged the unknowable. It reminded the world that humans can surmount seemingly impossible obstacles, while reinforcing a realization of our tiny place in the cosmos. At the end of a year marked by international strife, violence, and conflict, the human race came together in awe and reverence, significantly at a time when many were also surrounded by family and loved ones at home. And as families gathered together around the world, all were also mindful and concerned for the safety of the three voyagers a very long way from home.

For NASA and the United States this was an unprecedented marketing and public relations triumph. After a year of front pages displaying images of carnage in Vietnam, and assassinations, racial unrest, bloody street riots, and student protest at home, America had done something that had lifted the spirits of everyone alive. The following day, the press led with stories reporting how this one event solidified tremendous international goodwill for the entire American people. Secondarily, this was also the decisive turning point when the world press was forced to acknowledge that America had dramatically pulled ahead of the Soviet Union in the race to the Moon.

There were a few derisive voices. A small minority of commentators questioned what they perceived as the inappropriate inclusion of a religious text in a secular mission funded by the American taxpayers. And notably, atheist activist Madalyn

Murray O'Hair filed a highly publicized lawsuit against NASA, seeking to prohibit astronauts from any future public prayer in space as a violation of the separation of church and state. The lawsuit was eventually rejected by the U.S. Supreme Court, but made NASA cautious to avoid another such incident. Not that the majority of Americans had such reservations; in polls the American public broadly supported the reading from Genesis. And for some, it was a suitable rejoinder to America's archrivals, the "Godless Communists."

Above: *Apollo 9 lunar module in Earth orbit.*

Below: *The Westinghouse Lunar Surface Black-and-White Camera was made famous when it recorded what is, perhaps, the most iconic live television moment in history: Neil Armstrong's descent down the lunar module ladder and his first steps on the lunar surface during Apollo 11. Armstrong activated the camera by pulling a lever at the "porch" at the top of the ladder.*

Apollo 9

In Earth orbit, Apollo 9 was the first flight test of the lunar module. The many systems to be used in the pending lunar landing mission needed to be checked out,

including the LM spacecraft, the capability of ground tracking stations to gather telemetry from two spacecraft simultaneously, the Portable Life Support System backpack to be used by the astronauts on the lunar surface, and the lunar surface camera system. (On board *Spider,* the Apollo 9 lunar module, was a new Westinghouse lunar surface camera. No television camera flew aboard the command module, *Gumdrop*.)

The black-and-white Westinghouse lunar surface camera was a true engineering marvel and famously transmitted all images from the Moon on Apollo 11. Westinghouse was given this specific contract as a result of important advances they had made developing a top-secret Secondary Electron Conduction (SEC) video tube for the U.S. Department of Defense. Invented to provide enhanced military surveillance under conditions of restricted lighting—a twilight search for a downed pilot in the jungles of Vietnam, for instance—the highly classified SEC tube was both small and very effective at capturing images under low-light conditions. Logically, Westinghouse and NASA also suspected the SEC tube might also be ideal for shooting television from the lunar surface.

The challenges facing the Westinghouse engineering team were unique. This complex and intricate device needed to be both portable and easy to operate by astronauts wearing pressure suits; it had to be able to withstanding the massive G forces of a Saturn V liftoff; and it had to remain fully operational under the extreme temperatures of the lunar environment, which ranged from +250 degrees Fahrenheit in direct sunlight to -250 degrees Fahrenheit in the shade. Like the previous RCA camera flown on Apollo 7 and 8, the Westinghouse used the earlier standard: the slow-scan system of 10 frames per second, with 320 lines per frame.

But the Westinghouse engineers reported that the two

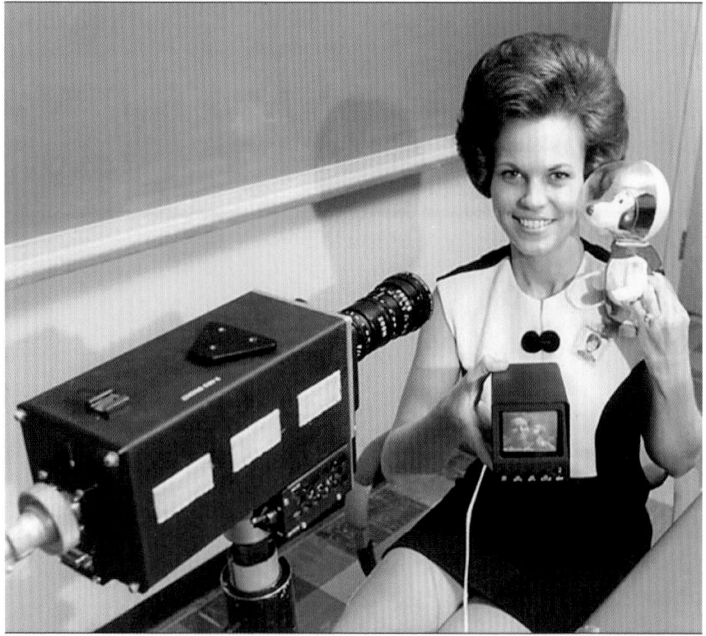

LEFT: Astronaut Eugene A. Cernan points to Apollo Landing Site 2 on a lunar map in this image captured from the fifth telecast made by the color television camera aboard the Apollo 10 spacecraft. When this picture was made, the Apollo 10 spacecraft was approximately 175,300 nautical miles from Earth, and only 43,650 nautical miles from the Moon. Cernan was the Apollo 10 lunar module pilot.

RIGHT: The Westinghouse Color Camera with its tiny video monitor first flown on Apollo 10. "Snoopy" was the name of the lunar module for this flight.

OPPOSITE PAGE: Internal color wheel of the Westinghouse color camera.

toughest parts of their assignment were reducing the raw weight of the camera and creating a reliable system that could operate on very little power. During the 1960s, microelectronics was still in its infancy, and the Westinghouse team used forty-three integrated circuits in its design to reduce weight and increase the energy efficiency. The final result weighed only seven pounds and operated on a mere seven watts, the energy comparable to illuminate a 1969-era Christmas tree bulb.[14]

Even though television transmissions during Apollo 9 were fairly limited and confined to the interior of the lunar module—a live telecast during astronaut Rusty Schweickart's EVA outside the LM was canceled—the Westinghouse produced good results.

Apollo 10

A dress rehearsal flight, Apollo 10 took the command module and lunar module around the Moon and into orbit where the

spacecraft practiced maneuvers that would be performed during a full lunar landing, climaxing with the LM's descent to within 50,000 feet of the Moon's surface. At commander Tom Stafford's urging, Apollo 10 was also the first to tryout the Westinghouse color camera. It came with both a zoom lens and a tiny external monitor, which enabled the crew to see the images they were sending home. (Prior to this flight, properly aiming the camera often required listening to positioning directions from Mission Control.)

The prevailing color television standards of the late 1960s required cameras that were far too large and too heavy for practical use on the Apollo missions. But by late 1968, Westinghouse, which had been moving ahead with their lunar surface black-and-white camera, was stealthily experimenting with a lightweight color television alternative, a single-tube color camera using an abandoned technology dating back to 1950. (It was developed by the renowned CBS Laboratory engineer Peter Goldmark, the pioneer responsible for the once ubiquitous 33⅓ rpm long-playing record.) The 1950 CBS color system employed an internal spinning wheel that allowed a single video pick-up tube to sequentially scan red, green, and blue images that could be seamlessly recombined into a single color image.)

The first laboratory tests were conducted in February 1969 and the results were so promising that Westinghouse hoped to get this on an Apollo flight later that year. However, a major obstacle remained: even though the clever use of an obsolete technology reduced the weight and power of the television equipment on board the spacecraft, the television signal from

14. 29. Westinghouse Program Manager for the Apollo Lunar Camera Stan Lebar speaking at a July 16, 2009, news briefing and symposium held at the Newseum, Washington, D.C.

Apollo would be unsupported by the world's telecommunications infrastructure. So Westinghouse immediately began to design conversion equipment capable of taking live color-wheel scan transmissions and making them immediately compatible with American NTSC television standards. The camera and its associated cabling weighed 8.5 pounds. When commander Tom Stafford saw an early test of the new camera he insisted that it be included on the Apollo 10 mission.

Four hours into the flight, as the command module *Charlie Brown* was executing a docking maneuver with the lunar module *Snoopy*, Stafford was eager to try out the new camera. "I took this opportunity to fire up our TV camera, giving the world their first color look at the Earth falling away beneath us," he wrote. "This was the first live color television picture ever transmitted from space. And our little skunk-works camera was so good you could see the rivets on *Snoopy*'s metal skin as we closed in for the docking."[15]

Stafford had a natural talent for making the broadcasts lively and engaging to the audience back home. During a television broadcast showing the Earth receding in space, he said: "You can tell the members of the British Flat Earth Society that they are wrong: The Earth is round." Their president had a message the next day: "Colonel Stafford, it may be round, but it's still flat, like a disk." Later in the mission, he treated viewers to the first live images of a crew member shaving in space. Such intimate and playful gestures on this flight and many that followed did much to engender audience identification and empathy with the Apollo voyagers, which, in turn, had tremendous public relations value for both NASA and the country.

Apollo 11

Without doubt, live television played its greatest role during the mission that fulfilled John F. Kennedy's goal to put a man on the Moon by the end of the decade. A version of the Westinghouse lunar camera that had been tested previously on Apollo 9 was onboard the LM *Eagle*, stored in the Modularized Equipment Stowage Assembly (MESA) in the descent stage of the vehicle. It waited there, ready to be deployed as Neil Armstrong emerged from the LM's front hatch. Indeed, the mission transcript from the five minutes immediately preceding the historic first step indicate how important the live television images were to the mission.

BUZZ ALDRIN (*lunar module pilot*): Did you get the MESA out?
NEIL ARMSTRONG (*commander*): I'm going to pull it now.
NEIL ARMSTRONG: Houston, the MESA came down all right.
BRUCE MCCANDLESS (*capsule communicator—"capcom"—at the Manned Spacecraft Center in Houston*): This is Houston, Roger. We copy. And we're standing by for your TV.
ARMSTRONG: Houston, this is Neil. Radio check.
MCCANDLESS: Neil, this is Houston. Loud and clear. Break. Break. Buzz, this is Houston. Radio check, and verify TV circuit breaker in.
ALDRIN: Roger, TV circuit breaker's in, and read you five square.
MCCANDLESS: Roger. We're getting a picture on the TV.
ALDRIN: You got a good picture, huh?
MCCANDLESS: There's a great deal of contrast in it, and currently it's upside down on our monitor, but we can make out a fair amount of detail.
ALDRIN: Okay. Will you verify the position—the opening I ought to have on the camera?
MCCANDLESS: Stand by.
MCCANDLESS: Okay. Neil, we can see you coming down the

15. Stafford and Cassutt, op. cit.

LEFT: Buzz Aldrin descending the ladder from the lunar module.

RIGHT: The deployment of the flag of the United States on the surface of the Moon is captured on a frame of 16mm film during the Apollo 11 lunar landing mission. Here, astronaut Neil A. Armstrong, commander, stands at the left of the flag's staff. The image comes from film exposed one frame per second by the time-lapse 16mm Data Acquisition Camera (DAC)—aka the sequence camera — mounted in the LM pilot's window. Below it are images from the final broadcast, July 23, 1969, of the Apollo 11 astronauts, Buzz Aldrin, Neil Armstrong, and Michael Collins.

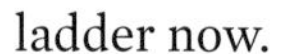

ladder now.

ARMSTRONG: Okay. I just checked getting back up to that first step, Buzz. It's—not even collapsed too far, but it's adequate to get back up.

MCCANDLESS: Roger. We copy.

ARMSTRONG: It takes a pretty good little jump.

MCCANDLESS: Buzz, this is Houston. F/2—1/160th second for shadow photography on the sequence camera.

ALDRIN: Okay.

ARMSTRONG: I'm at the foot of the ladder. The LM foot pads are only depressed in the surface about one or two inches, although the surface appears to be very, very fine grained, as you get close to it. It's almost like a powder. Down there, it's very fine.

ARMSTRONG: I'm going to step off the LM now.

ARMSTRONG: That's one small step for man, one giant leap for mankind.

It was 10:56 P.M. in Washington and New York, a muggy Sunday evening on most of the East Coast. At this point in the live television broadcast, NASA and the television networks superimposed text on the screen for the world to see: "Live from the surface of the Moon" or "Armstrong on Moon." Forty years later, Neil Armstrong admitted that no member of the human race was more surprised than he when he heard from Bruce McCandless at Mission Control that they were looking at a live television picture from the Moon.[16]

After Aldrin emerged from the LM a few minutes after Armstrong, the camera was placed on a tripod several dozen yards away connected by a 100-foot cable. The world watched the ghostly live images of Neil Armstrong and Buzz Aldrin on the lunar surface with the LM in the background. During their single extra-vehicular activity (EVA, commonly known as a "moonwalk"), the pair raised the American flag, unveiled a commemorative plaque, took a phone call from President Nixon seated in the Oval Office, deployed a small array of scientific experiments, and collected soil and rock samples. The entire EVA broadcast lasted a little more than two hours.

Unfortunately, the live television from Apollo 11 that the world saw in 1969 did not feature the best possible images. The conversion of the slow-scan S-band television telemetry received from the Moon into a format compatible for international television viewers resulted in a degraded image, which was further eroded as the signal was sent via satellite and cables throughout the world.

One of the Westinghouse color television cameras, previously flown on Apollo 10, was aboard the Apollo 11 command module *Columbia*, where it was used for broadcasts both on the way to the Moon and during the return to Earth. During the first of their broadcasts on the return voyage, command module pilot Mike Collins conducted some zero-G demonstrations including holding a spoon of water upside-down to show that it did not spill. And in their concluding transmission, on July 23, Armstrong, Aldrin, and Collins offered their reflections about what the mission meant to them personally.

Apollo 12

Apollo 12 was notable as the first mission to have a live color camera on the lunar surface, a modification of the Westinghouse camera used on Apollo 10. Once again stored in the LM descent-stage MESA, the color camera was deployed by commander Pete Conrad just before his trip down the ladder of the

LM *Intrepid*. For the first hour of the first EVA, the technology worked brilliantly. And then came the moment that tragically demonstrated the incalculable value of having a live television camera on the Apollo lunar missions. As lunar module pilot Alan Bean was positioning the camera, the lens was accidentally pointed directly at the unfiltered sun, burning out the SEC image processor and abruptly ending all television for the rest of Apollo 12's stay on the lunar surface.

The situation was entirely avoidable. NASA's astronaut training never focused on how to effectively use the television cameras or on television's potential for building and enhancing public interest and support. In post-mission debriefings, the crew noted that they were never properly trained in the use of the television equipment. Despite winning the battle to get television onto Apollo, it was relegated to a low priority in the hierarchy of tasks. Yet, in hindsight, the lasting images the lunar astronauts bequeathed to future generations are indisputably among Apollo's most important legacies.

When the live lunar television abruptly stopped, American commercial networks were immediately forced to rely on emergency back-up plans. CBS News broadcast pictures of two space-suited men carrying out the planned EVA timeline in the Bethpage, Long Island, LM studio, while the actual live voices from the Moon occupied the audio. At NBC, the alternative was even more unorthodox: the network called upon the talents of famed puppeteer Bil Baird and his assistants to enact the EVA using marionettes constructed months earlier in Baird's Greenwich Village workshop and theater. "Afterwards we were amazed to learn that many watching the telecast didn't know the two men in space suits weren't real," Baird said in a contemporary news article. "NBC projected the word 'simulation' on the screen at all times, but it seems most TV watchers don't know what it means."[17] Doug Ward, who worked in the Public Affairs department during Apollo, contends the Apollo 12 mishap irreparably affected all future network television coverage. "Without live television, there wasn't much of a story for the networks; they all cut away from it. And once they found that they could get away with that, it became the standard for covering the rest of the missions. So the loss of that camera on Apollo 12 was almost as big a factor

16. Armstrong quoted in a letter read by Stan Lebar, Westinghouse Apollo Lunar Landing TV Program Manager, at a July 2009 symposium held at the Newseum, Washington, D.C.
17. "Bil Baird Conquers Earth and Space," by Norman Nadel, syndicated new story.

ABOVE: *This unusual photograph, taken during the second Apollo 12 EVA, shows two U.S. spacecraft on the surface of the Moon. The Apollo 12 lunar module is in the background, the unmanned Surveyor 3 in the foreground. Surveyor 3 soft-landed on the Moon on April 19, 1967. The Apollo 12 LM, with astronauts Pete Conrad and Alan Bean aboard, landed about 600 feet from Surveyor 3 in the Ocean of Storms two-and-a-half years later, on November 19, 1969. The television camera and several other pieces were removed from Surveyor 3 and brought back to Earth for scientific examination. Here, Conrad examines the Surveyor's TV camera prior to detaching it.*

BELOW: *One of the few existing photographs that show the Bil Baird marionettes during NBC's live network coverage of the Apollo 12 moonwalks. The three major networks had to call on emergency back-up plans when the lunar television camera failed an hour into the first moonwalk. According to a November 21, 1969, article in the* New York Times, *in comparison to the live full-scale simulations seen on CBS and ABC, "NBC's puppets were less successful. They were manipulated from an overhead bridge on a scale model of the Moon landscape. Three men worked two puppets. . . . The NBC puppets, while slightly awkward, were turned into an advantage by the network . . . [when they occasionally] relieved the monotony."*

as anything that happened during the Apollo program to dictate a reduction in television network coverage."[18]

Not surprisingly, the number of viewers staying with the coverage was extremely low. And this wasn't the only factor affecting public interest in Apollo 12. Unlike Apollo 11, which touched down at 4:17 P.M. EDT on a Sunday afternoon when many were at home, the landing of Apollo 12 came at 1:54 A.M. EST during the middle of the workweek. (The two EVAs took place outside of prime time: the first during the morning hours in the United States; the second when much of the country had already gone to bed.)

Other matters were also distracting the attention of the American public. Apollo 12's mission was fatefully destined to coincide with two other developing news stories, which garnered banner headlines during much of the flight. The day before liftoff, Vice-President Spiro T. Agnew delivered an attack against the "unelected" elite media, accusing the major television news departments of bias and distortion, a broadside that proved to be a pivotal moment in the history of the Nixon administration and its relationship with the press. On November 20, 1969, the day *Intrepid* lifted off from the Moon and rendezvoused with CM *Yankee Clipper*, shocking and explicit photographic evidence of the My Lai Massacre was published for the first time, just a week after reporter Seymour Hersh broke the story of the notorious 1968 Vietnam atrocity.

After the triumph of Apollo 11, the country had moved on. For more than a decade, Americans understood "the space race" as a Cold War contest in which pride and national security were equally at stake. After the goal had been achieved—flawlessly and for the entire world to witness—public interest turned elsewhere. The reduced network coverage of Apollo 12 reflected not only the budgetary concerns of the television news departments, but the national mood as well. November 1969 marked the beginning of the decline in commitment and resources covering Apollo by all three television networks.

Apollo 13

Determined that the camera mishap on Apollo 12 would never reoccur, NASA made certain that a lens cap and a securing lanyard were added to the Westinghouse lunar color camera on the next flight. (Also on board was a backup black-and-white lunar camera similar to the one flown on Apollo 11.) But a far more dangerous mishap would define the lasting legacy of Apollo 13.

During the opening seconds of what appeared to be an otherwise routine mission, NASA accomplished a television first when a color television camera was placed at the 360-foot level of the launch tower to capture live images of the fiery lift-

18. Doug Ward, interview with the authors, November 29, 2011 and December 12, 2011.

off. Westinghouse produced a special shielded sequential color camera, which employed a burn-proof image sensor and heat-sink mesh designed to withstand temperatures estimated to be in the range of 2 million degrees Fahrenheit.

During the long coast to the Moon, the crew conducted three lengthy television transmissions, two occurring during the television networks' prime time programming. Unbeknownst to the astronauts on Apollo 13, however, the networks had stopped carrying the live broadcasts from space. It was shortly after the conclusion of a television transmission on the evening of Monday, March 13, that an explosion occurred in an oxygen tank, crippling the spacecraft. At first, only the reporters covering Mission Control had an indication of the serious nature of the situation. But around midnight, prominent newsmen, including CBS News's Walter Cronkite, were rushed on the air as the networks cut away from their late-night talk shows to carry a live news conference from Houston.

Suddenly, television was important again. Astronaut training required preparing for hundreds of potential crises, but no one had ever foreseen this particular scenario. Television was covering the emergency live as it unfolded, and some commentators openly expressed doubt about the possible success of the return mission. Though there would be no more television transmissions from the spacecraft, the networks repeatedly cut into regular programming with news updates, and covered the LM engine burn near the Moon and every mid-course correction as major news events. People throughout the world were transfixed on the fate of the three voyagers.

On Friday, April 17, 1970, the command module *Odyssey* separated from the service module and then from their lunar module lifeboat *Aquarius,* and prepared for a precarious reentry unlike any that had ever been attempted. In New York City's Grand Central Station, commuters watched a large screen television broadcast, just as millions of others across the globe nervously followed the final minutes of the ill-fated mission. In one of the most dramatic returns from space ever witnessed on live television, cameras on the USS *Iwo Jima* caught vivid images of *Odyssey* as the three red and white striped parachutes deployed for a near pinpoint splashdown in the Pacific.

TOP: The CBS Evening News with Walter Cronkite followed the perilous journey of Apollo 13 as it unfolded.

MIDDLE: Overall view of the crowded Mission Operations Control Room (MOCR) in the Mission Control Center (MCC) at the Manned Spacecraft Center (MSC) during post-recovery ceremonies aboard the USS Iwo Jima, *the prime recovery ship for the Apollo 13 mission. The Apollo 13 spacecraft, with astronauts Jim Lovell Jr., commander; Jack Swigert, command module pilot; and Fred Haise, lunar module pilot, aboard, splashed down in the South Pacific at 12:07:44 P.M. CST, April 17, 1970. The smooth splashdown and recovery operations brought an end to a perilous spaceflight for the crew members, and a tiring one for ground crew in MCC.*

BOTTOM: When the explosion on Apollo 13 turned a routine mission into a crisis, all live television from the spacecrafts was halted. Television audiences were riveted to the coverage, with live reports occurring throughout the tense days. The splashdown in the Pacific Ocean on April 17, 1970, was broadcast live via cameras on the USS Iwo Jima. *This image shows the command module* Odyssey *being lifted onto the ship soon after splashdown.*

Astronaut Alan Shepard, Apollo 14 commander, can be seen preparing to swing at a golf ball during a television transmission near the close of EVA-2 at the Apollo 14 Fra Mauro landing site. Shepard is using a real golf ball and an actual six-iron, attached to the end of the handle of the contingency sample scoop. Astronaut Ed Mitchell, lunar module pilot, looks on. Also visible in the picture is the erectable S-band antenna (left foreground).

Apollo 14

The next flight, nine months later, fulfilled the aborted flight plan of Apollo 13. Besides the addition of the lens cap and

back-up black-and-white camera, the Westinghouse color lunar camera had been customized with a new "burn-proof" video sensor. But the color camera's limitations were also evident on this flight. Mounted on a tripod and tethered to the LM by its cable, the camera often transmitted little more than an unchanging static image of the lunar surface while commander Alan Shepard and LM pilot Ed Mitchell conducted their explorations out of the lens's line of sight. This made for tedious television, and the networks received complaints from viewers angered by the preemption of their favorite daytime programs. A secondary S-band antenna deployed during the EVA to handle additional television bandwidth improved the picture quality slightly, but many video engineers remained convinced there was much room for improvement.

The television highlight of the flight came near the end of the second EVA, when Alan Shepard stood before the camera and made golf history. Shepard attached the customized head of a Wilson six-iron to a lunar-sample scoop handle to drive several golf balls. His large and awkward lunar suit required a one-handed swing but he appeared to hit the second shot well, saying it went "miles and miles and miles" in lunar gravity, which is one-sixth that of Earth's.

Apollo 15

The ghostly video transmissions from the first three Apollo expeditions to the lunar surface had astounded nearly everyone.

Yet some NASA engineers and video specialists were convinced that better image resolution was possible.

By early 1969, RCA's pioneering black-and-white Apollo CM television camera had been replaced by the more sophisticated models developed by their rivals at Westinghouse. Not surprisingly, RCA was eager to once again become Apollo's chosen television contractor. Like their competitor, RCA was also a leading manufacturer of consumer television sets, a brand familiar to nearly every American considering the purchase of a new living room console. To be chosen by NASA as the manufacturer of choice conferred a prestige endorsement and repeated brand-name exposure in the extended news coverage.

RCA's chance to reclaim its former status came as the Apollo lunar program entered its mature stage with the final three extended-duration "J" missions. As part of the planning of these later missions, NASA was given an ideal opportunity to rethink live television's role and its potential. The old system, with a camera tethered to the lunar module via a lengthy extension cord, wouldn't work on the J missions, during which the use of the new lunar roving vehicle (LRV) would transport the two astronauts long distances and out of visual range of their base camp.

By late 1969, NASA administrators began discussions about designing a self-contained mobile independent television system mounted on the rover, sometimes referred to as the "moon buggy." Simultaneously, RCA had under development a Silicon Intensifier Target tube, a highly sensitive video receptor considered an advancement on the Westinghouse SEC sensor. A major difference was that the RCA tube could withstand accidental exposure to direct sunlight and not sustain permanent damage. (A second lunar television mishap, like the one on Apollo 12, would be a public relations disaster.) Additionally, the Silicon Intensifier Target tube was more versatile: it could reproduce subtle gradations within deeply shadowed areas yet didn't display any lingering image carryover, a major cause of the shifting, ghostly appearance of the earlier Apollo lunar telecasts.[19]

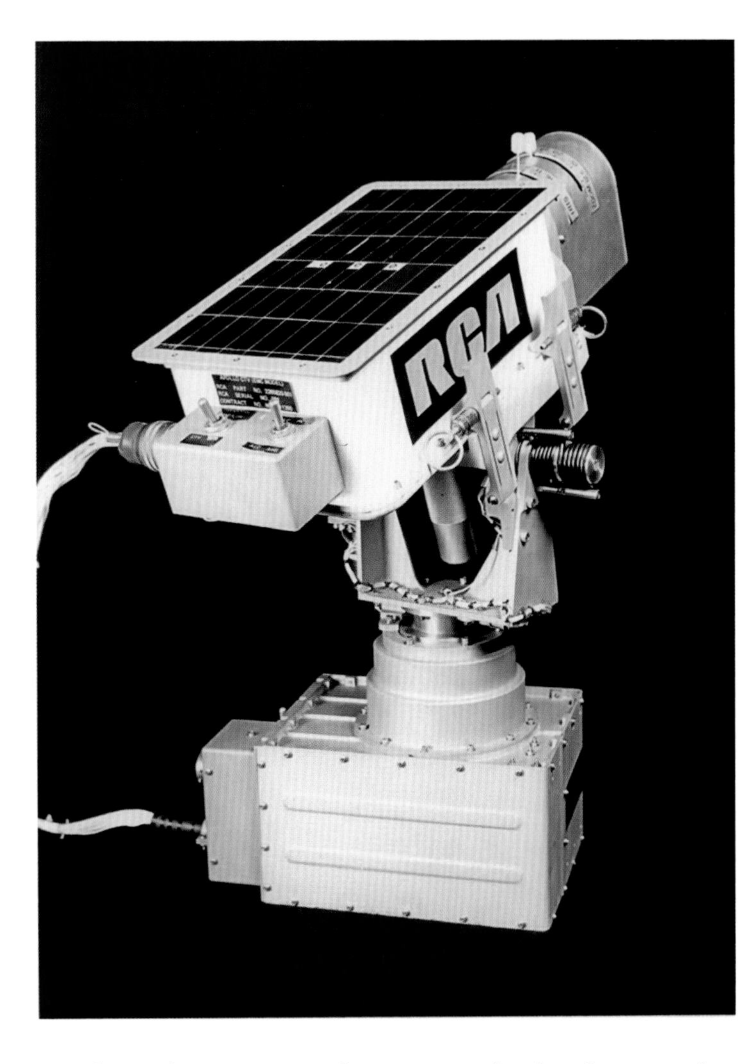

An early prototype of the RCA GCTA camera used on the final three Apollo lunar rover missions. Installed on the rover, the GCTA was covered in gold thermal insulation.

In late July 1970, NASA chose RCA to develop the Grounded Command Television Assembly (GCTA)—better known as "Gotcha" to the RCA team—the lunar rover's own television camera, which was designed to respond to remote commands from an operator at the Johnson Space Center. Rather than interrupting the moonwalk with verbal requests to the astronauts to pan the camera to a particular landscape feature or zoom in on a specific rock in the foreground, an operator at Mission Control—working in conference with scientists in Houston—would act as an independent exploratory third eye while also following the progress of the moonwalkers.

Like the earlier Westinghouse Apollo color television camera, the new RCA camera also utilized the old CBS Laboratory spinning color wheel technology. In conjunction with the GCTA, RCA also built the Lunar Communications Relay Unit (LCRU), a suitcase-size array of instruments that routed communications signals from the LRV to the Earth via a 28-inch diameter, gold wire-mesh, high-gain dish antenna mounted on the front of the rover.

The improved image quality was immediately apparent to anyone following the television coverage of Apollo 15. At each stop on the lunar rover journey, the astronauts would engage the camera and manually point the high gain antenna at the Earth. The RCA camera's second major revelation came a half-hour later when it was attached to the GCTA and the lunar communications relay unit on the lunar rover. Then, Ed Fendell at the Manned Spacecraft Center took control of the camera's movement, using a small panel containing only eighteen push buttons. He could pan the image left or pan it right; he could zoom in or zoom out; he could adjust the angle up or down; and he could control the brightness of the image. A communications specialist at NASA since 1964, Fendell was instrumental in the development of the GCTA system and suddenly became the subject of news profiles. He was nicknamed "Captain Video" by the press.

When the rover came to a stop, Fendell would conduct a slow camera pan of the area, while an assistant would record the panorama in segments using an instant camera to compile a crude mosaic photographic image that could be referenced for immediate orientation. Fendell was simultaneously in communication with NASA scientists, who would pass on requests. Should a geologist want a closer look at a particular boulder on the horizon, the GCTA could be commanded to zoom in and observe it in better detail.

However, operating the remote lunar rover camera with those eighteen push buttons proved to be an intense assignment. In a 2000 interview, Fendell described the experience. "You really had to pay attention. You couldn't stop and think, and you couldn't stop and go to the bathroom, and you couldn't do too much talking with anybody else, but you had to monitor the other [communications] systems, too. . . . If you went through six hours of that, you were really tired. It really wore you down, because it was constant. The only time you got to take a breath is when [the camera wasn't transmitting while the astronauts] started to move, and they took over the rover, and that was your break time, like, 'I'm going to the bathroom, quick. I'll be back.'"[20]

19. Sam Russell, *Shooting the Apollo Moonwalks: A Recollection of How It Was Done*, NASA, no date.
20. Ed Fendell, Johnson Space Center Oral History Project, Oral History Transcript, October 19, 2000, p. 60.

LEFT: Apollo 15 commander Dave Scott confirms Galileo's hypothesis that, in the absence of air resistance, all objects fall with the same velocity. A geologic hammer in Scott's right hand and a falcon feather in his left hand reached the surface of the Moon at the same time. The demonstration was performed before the television camera on the lunar roving vehicle; no still photographs were made of this event.

RIGHT: The color television camera and self-contained broadcast station mounted on the Apollo 15, 16, and 17 lunar rovers was used to broadcast the moment that the ascent stage of the lunar module lifted off. The rover was parked nearby and, in a technological triumph for NASA—and for Ed Fendell, who controlled the camera from his console at the Manned Spacecraft Center in Houston—the camera was able to follow the liftoff and show viewers the ascent as it happened.

Naturally, Fendell's primary task was to train the camera on the moonwalkers, monitoring their progress as they worked, zooming in when they were active in one place, or providing pans and wide angles to show a location. The astronaut's commentary provided the soundtrack to the live pictures, creating one of the most entertaining and absorbing extended looks at men at work ever broadcast on live television.

"The thing that made the lunar surface TV work so well was that it was controlled from the ground and it did exactly what you'd want a TV camera to do: show the astronauts and their activities without any participation required on their part," NASA's Doug Ward said in retrospect. "The astronauts could forget about it and they did, and it was just a beautiful way for the public to watch over the shoulder and to absorb all the activities on the Moon."[21]

On Apollo 15, commander Dave Scott even performed a live science demonstration for the television cameras. At the conclusion of his third EVA, in one hand Scott held the geologic hammer he had used on the lunar surface and in the other hand a falcon feather (the Apollo 15 LM was named *Falcon*). And on his verbal cue Scott dropped both at the same time proving Galileo's 1638 hypothesis, which proposed that objects of unequal weights would fall at the same speed in a vacuum. Scott's experiment, captured live, has lasted the test of time. Generations later, this video is often used in science classrooms around the world. These human-interest television moments are among the most remembered and replayed of the entire Apollo program, demonstrating the power of television to tell a story.

The GCTA also provided the world with its first look at a lunar liftoff as *Falcon* fired its engine to rendezvous with the CM *Endeavour*. Because of a problem with a jammed clutch on the GCTA, the camera could not move upward to follow the LM as it rose away from the lunar surface (a problem that was resolved on the final two Apollo flights).

Dr. Frank N. Stanton, then President of CBS, was unusually vocal in his enthusiasm about this flight and its coverage. On August 9, 1971, two days after splashdown, in a memo addressed to the CBS Organization, he wrote: "We can all take pride in CBS's part in the Apollo 15 mission: the superb coverage of CBS News, and the color system of CBS Laboratories which made it possible to send pictures back live to the Earth. From every standpoint, our coverage of Apollo 15 was journalism of the highest order and broadcasting at its very best. To all of you who had a hand in making it possible, the rest of us who watched and listened in awe say, 'well done.'"[22]

Unfortunately, the elation and good cheer faded fairly quickly in the corner offices of the big three networks. The Apollo 15 moonwalks proved to be the last broadcast live in their entirety by CBS, NBC, and ABC. As Ed Fendell recalled in a 2000 interview, "On 15, the United States networks carried it all . . . in real time. We knocked the soaps off the air. It was prime time and so on. In Europe, it was the same. On 16, we didn't show . . . we could not move anybody out of the way."[23]

21. Doug Ward, interview with the authors, December 12, 2011.
22. Stanton's memo is quoted in Alfred Robert Hogan, "Televising the Space Age: A Descriptive Chronology of CBS News Special Coverage of Space Exploration from 1957 to 2003," Masters Thesis, University of Maryland, 2005; p. 245
23. Ed Fendell, Johnson Space Center Oral History Project, Oral History Transcript, October 19, 2000, pp. 23, 62.

Apollo 16

The penultimate voyage to the Moon was designed to build on the success of Apollo 15, with three more long-duration

lunar rover EVAs and refined and enhanced resolution television transmissions. The GCTA system remained unchanged, though the use of a larger 210-foot receiving dish on Earth boosted the signal's strength. In addition, NASA contracted with Image Transform,[24] a new visual technology firm in North Hollywood, to work from the live television signal and run it through their system to enhance, clean up and remove excess traces of visual "noise" before forwarding it on to Houston and the world television networks. The improved quality of the images was dramatic, but few in America saw them. Inside NASA there was growing concern about the shrinking national television audience, and increased indication the networks were losing interest. In response, Apollo 12 astronaut Alan Bean circulated a memo suggesting that NASA spokespeople—including experienced astronauts—might serve as commentators during the coming live broadcasts of the Apollo 16 moonwalks, offering insight and commentary while also injecting some soft marketing to explain how the Apollo missions benefit all Americans.[25]

Now convinced of the power of live television, NASA adjusted the flight schedule to position some of the EVAs during America's prime time viewing hours. It was assumed that by scheduling the moonwalks when Americans were home from work and school, NASA would regain some of the audience (and public support) that had fallen away since Apollo 11. A *Los Angeles Times* story published in 1971 speculated that NASA's hunger for high ratings was tied to concern about the agency's future; if congressmen suspected the public was uninterested, politicians would feel less reluctance to reduce appropriations for future space exploration.[26]

Ironically, NASA's EVA scheduling change may have actually exacerbated the situation; the three networks chose not to preempt their lucrative prime time programs with live broadcasts from the Moon. In the corporate network news divisions in New York, a number of veteran journalists, NBC correspondent Jim Hartz and CBS producer Robert Wussler in particular, argued passionately for extended live coverage, but they

A television image transmitted during the second Apollo 16 EVA, on April 22, 1972. In the foreground, lunar module pilot Charlie Duke inspects the RCA GTCA camera, while commander John Young works in the background. During this EVA, Young and Duke explored Stone Mountain and achieved the highest elevation above the LM of any Apollo mission.

were fighting a losing battle. There was a larger shift occurring within the networks' news divisions that year as the budgets for live news were greatly reduced. During the summer of 1972, the three networks cut down their extended live coverage of the Democratic and Republican conventions, events that in preceding years had been given gavel-to-gavel coverage.

For Apollo 16, the reduced network airtime was substantial. CBS News, which had been a ratings leader during its recent Apollo coverage, cut its live special coverage by nearly 43%. The comparison in coverage is instructive: For the Apollo 15 mission, CBS broadcast twenty-three-and-a-half hours of special coverage, of which the three EVAs comprised around eighteen hours. For Apollo 16, a mission with EVAs and mission highlights of nearly the same duration, CBS offered only thirteen-and-a-half broadcast hours. The EVAs on both missions occurred over long weekends: Saturday, Sunday, and Monday in the case of Apollo 15, and Friday, Saturday, and Sunday for Apollo 16. Stalwart Walter Cronkite even received a polite but firm protest regarding his network's reduced coverage during a live on-air interview with the wife of one of the Apollo 16 moonwalkers.[27]

That so few viewers saw the breathtakingly clear Apollo 16 television images live as they were received, must have been a particularly bittersweet experience for all the NASA officials and the veteran journalists who had spent years of their lives working on and covering Project Apollo. So rather than witness live commander John Young and LM pilot Charlie Duke slowly become dwarfed in size as they approached a huge boulder (nicknamed the "House Rock") or gasp with concern

24. Image Transform was founded in 1971 by John D. Lowry, who later started Lowry Digital Images, one of the film industry's leading film restoration houses. After restoring hundreds of old studio films for DVD release—including *Casablanca*, *Star Wars*, and *Gone With the Wind*—Lowry was asked by NASA in 2009 to restore the existing Apollo 11 television tapes.
25. Summary of "Suggestion for our TV support During Apollo 16," memo dated August 17, 1971, in *Live From the Moon*, p. 211.
26. Cited in Alfred Robert Hogan, op. cit.: Nicholas C. Chriss, "Apollo 16 Also Aiming at Nielsens," *Los Angeles Times* wire dispatch in *New York Post*, Wednesday, October 13, 1971.
27. Alfred Robert Hogan, op. cit., p. 138.

LEFT: *During the Apollo 17 mission, Jack Schmitt collects rock samples from a huge boulder near the Valley of Taurus-Littrow.*

RIGHT: *Barbara and Tracy Cernan watch as Gene Cernan rehearses the planting of the flag in preparation for Apollo 17.*

the moment Duke attempted and failed to complete an athletic jump (resulting in a fall directly upon his backpack), most Americans interested in following the mission had to content themselves with a selection of videotaped highlights within the nightly newscasts. The subliminal message conveyed by network television's coverage and the public's lack of interest was clear and consistent: in the course of less than three years, an achievement that, when first accomplished, was acknowledged as a monumental turning point in human history, was slowly reduced in scope, magnitude, and importance into something commonplace. It was no longer major news.

Apollo 17

Unbowed by the tepid network coverage of Apollo 16 in April, NASA again scheduled the three lunar EVAs of the final mission to the Moon to coincide—and to inevitably conflict—with the evening programming of the three major networks. The December 1972 mission of Apollo 17 promised yet another spectacular first: a primetime night launch of the Saturn V, the only major American manned mission launched into the darkness over Cape Kennedy.

Among space watchers there was great excitement anticipating the moment when the Saturn V would generate its own daylight—likely confusing the local wildlife with a false dawn—as it slowly ascended into the heavens. At CBS, the Saturn V night launch provoked concern about the havoc a live news broadcast would cause among the network's loyal viewers. The liftoff was scheduled to occur during the time slot reserved for one of CBS's top-rated series, *Medical Center*. According to internal network memos, the plan at CBS was to briefly break into the episode of *Medical Center*, show the liftoff live, and then resume regular programming as soon Apollo 17 was no longer visible to camera.[28]

Within NASA it appeared likely that, once again, the networks would be carrying a mere fraction of the live Apollo 17 broadcasts. In early November 1972, news reports surfaced that NASA was hoping the two-year-old Public Broadcasting Service would carry coverage of the upcoming flight in its entirety. In fact, Henry Loomis, the new president of the Corporation for Public Broadcasting, had tendered just such a proposal to the individual PBS stations, specifying that NASA would oversee production and cover expenses. Loomis, the former head of Voice of America, was a recent addition to the CPB board, one of a group of new board members specially selected by the White House to control an outlet the Nixon Administration considered an enemy. (In the wake of the Pentagon Papers disclosures of 1971, PBS was one of many networks, newspapers, and magazines that angered 1600 Pennsylvania Avenue. In particular, the White House targeted PBS for programming it considered unsympathetic or offensive.) As a

28. CBS memo quoted in Alfred Robert Hogan, op. cit., p. 138.

government agency, the Corporation for Public Broadcasting was directed to avoid any role in programming; all PBS network content was produced, created, or acquired by the consortium of individual independent stations around the country. So when Loomis's Apollo 17 suggestion surfaced, many PBS station managers sensed a government effort to erode their power, and publicly voiced their ire. One PBS official said the Apollo 17 proposal set a dangerous precedent, "like letting General Motors to underwrite and produce programs for public television on car safety."[29] Perhaps in a more innocent time, such heated passions might have been avoided. But after years of demonstrable information manipulation and deception regarding Vietnam, the very notion of a government agency producing programming for public television stations, funded in part by tax dollars, was a potential public relations nightmare. Quickly realizing the CPB/PBS imbroglio could grow into a much larger problem, NASA abandoned its proposal for PBS Apollo 17 television coverage and the controversy disappeared from the national news cycle almost immediately.

When launch day arrived in early December, again the three commercial networks controlled how much of the Apollo 17 mission the American public would see. Despite all of CBS's advance planning, fate upended the network's prime time schedule on Wednesday evening, December 6, 1972. A technical problem with the automatic sequencer halted the Apollo 17 countdown a mere thirty seconds before its scheduled 9:38 P.M. EST liftoff. Not knowing exactly when the Saturn V would depart, the networks were forced to fill more than two-and-a-half hours of airtime before the actual launch at 12:33 A.M. EST the next day. In addition to the stress experienced by the network anchors and journalists forced to improvise commentary and background features while the countdown clock remained frozen at T minus thirty seconds, there can be little doubt that the network executives overseeing the prime time entertainment schedule did not remember that night fondly. The interrupted episode of *Medical Center* was never rejoined; the conclusion of its plot line was deemed important enough to be summarized for fans the following morning on the *CBS Morning News*.[30]

Viewers looking forward to watching extended telecasts of the EVAs were frustrated to discover the three networks offered little more than taped summary highlights, often delayed until 11:30 P.M. EST. After the extended coverage of the delayed launch, CBS's remaining Apollo 17 coverage totaled little more than an additional six hours of broadcast time, including the live splashdown.

When networks chose to cover the final moonwalks live, producers tried to retain viewers' attention with special features. On the *Today Show*, Jim Hartz interviewed Gene Cernan's nine-year-old daughter, Tracy. Later, when he viewed a video, Cernan reflected on the power of that moment. "There was a television screen behind her at the precise moment we were walking on the Moon. Hartz asked her questions, and one I'll never forget was: 'Do you think your daddy will find water on the Moon?' And she answered: 'Well, if he did, he went to the wrong place.' You cannot imagine how many parents and kids related to that. If you can get a kid's attention and make learning fun, you can teach them anything. That's the reward I get even today."[31]

When the third and final moonwalk took place during the live half-hour evening network news broadcasts on December 13, 1972, Apollo 17 wasn't the leading story on either NBC or CBS. Rather, within the newscasts there appeared five-minute summary reports featuring prerecorded excerpts. John Chancellor on NBC ended his segment with the words: "NBC News will continue to cover the Apollo mission this evening. . . . We'll break into *The Tonight Show* later this evening to show you Cernan and Schmitt climbing back into the lunar module. They very well may be the last men to walk on the Moon this century and we want to see their final actions."

Ironically, in one city, a single station decided to carry all the Apollo 17 moonwalks live. Independently, KUHT-TV, Houston's PBS station broadcast more than twenty-five hours of special reports. While the lunar rover was traveling and not transmitting pictures, KUHT supplemented their low-budget coverage with commentary from NASA officials and university geologists present in their Houston studio, who also answered call-in questions from viewers.[32] But then again, for Houston, this was a hometown event. ⦿

29. James Day, *The Vanishing Vision: The Inside Story of Public Television*, 1995, p. 233. Day suggests that the quotation, which appeared in the November 12, 1972, *New York Times*, was probably given by Jim Lehrer, then PBS coordinator of public affairs programming.
30. Alfred Robert Hogan, op. cit., p. 139.
31. Gene Cernan, interview with the authors, February 24, 2012.
32. Alfred Robert Hogan, op. cit., p. 58; television column in *The Baytown Sun*, (Baytown, Texas) December 8, 1972.

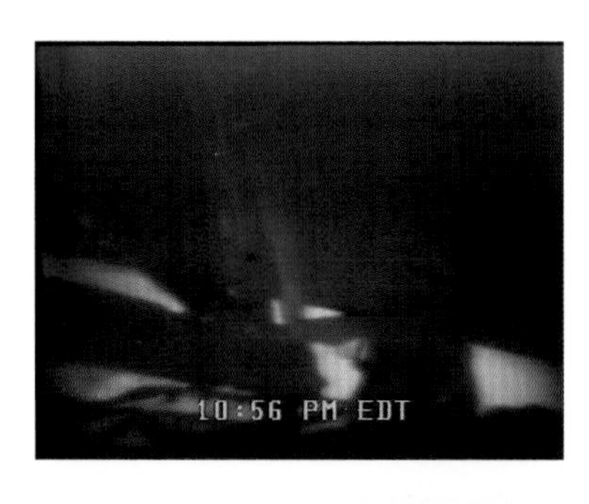
10:56 PM EDT

ARMSTRONG
ON
MOON

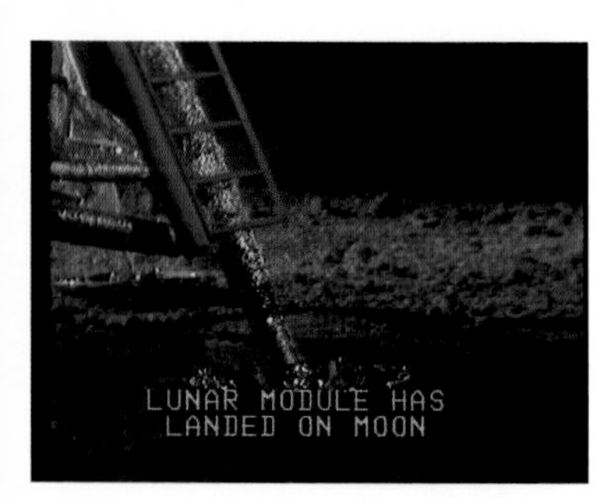
LUNAR MODULE HAS
LANDED ON MOON

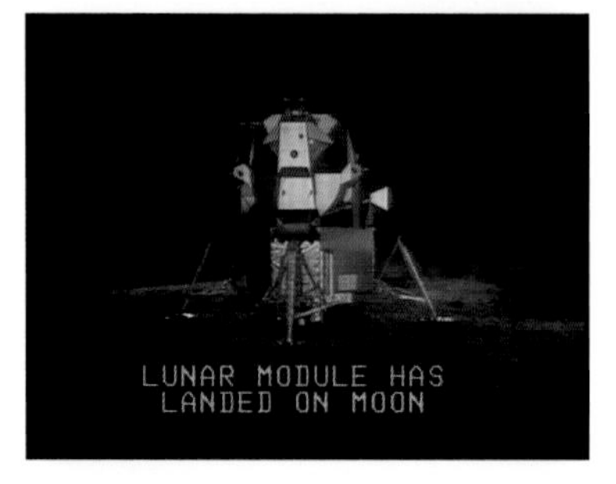
LUNAR MODULE HAS
LANDED ON MOON

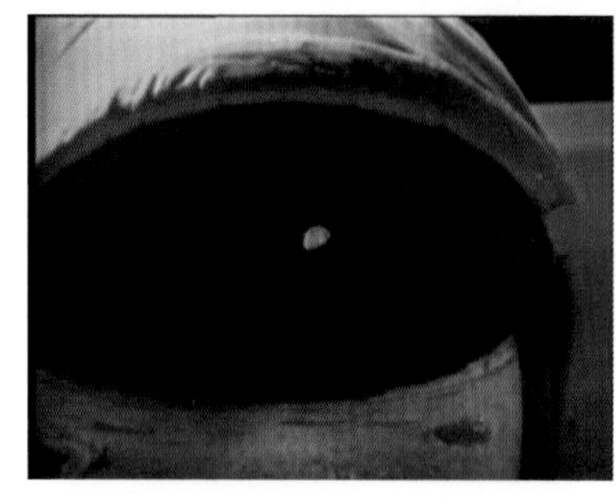

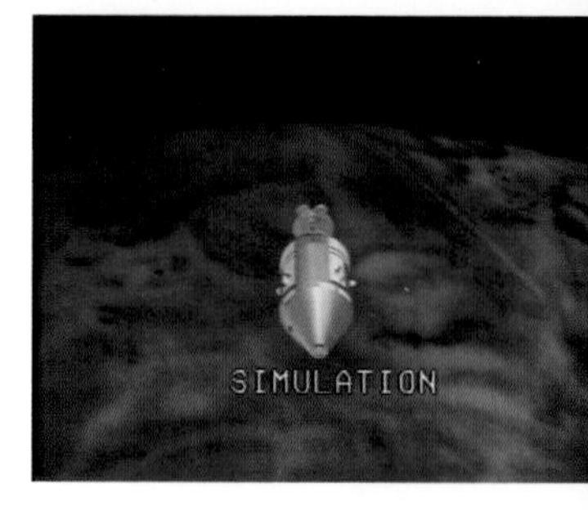
SIMULATION

CBS NEWS
SIMULATION

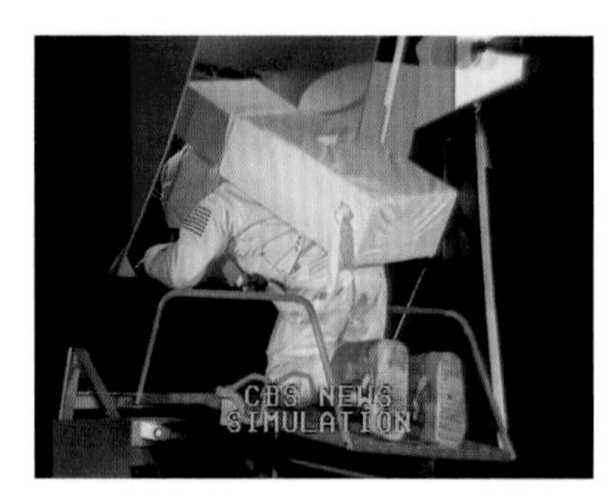
CBS NEWS
SIMULATION

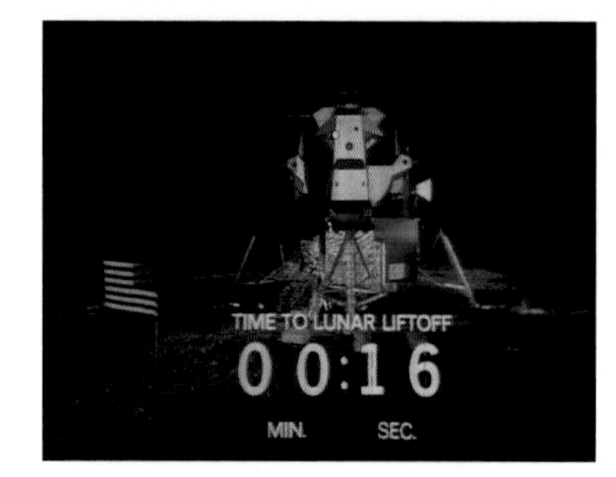
TIME TO LUNAR LIFTOFF
00:16
MIN.
SEC.

CBS NEWS
SIMULATION

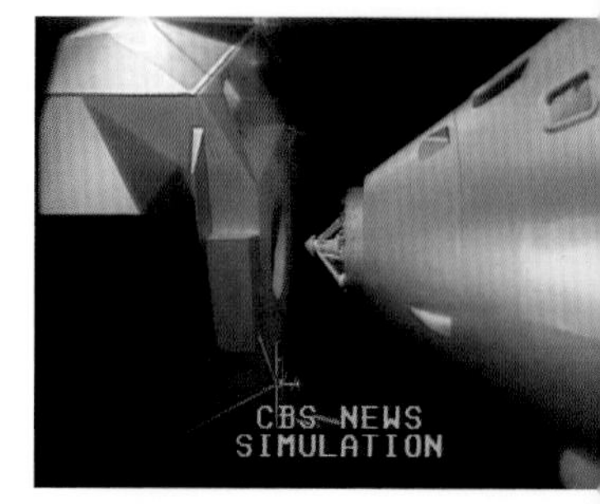
CBS NEWS
SIMULATION

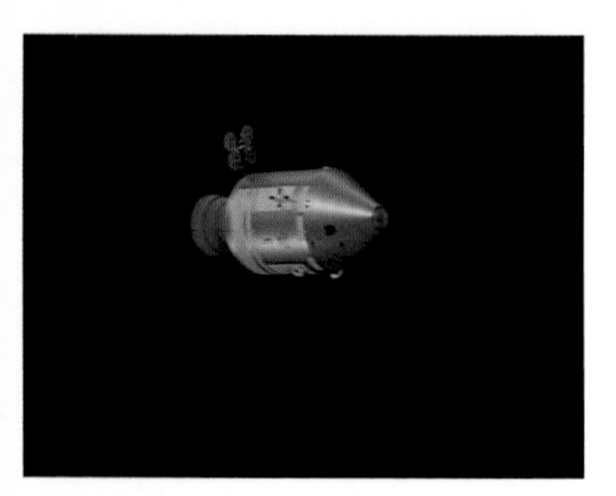

SIMULATION

LUNAR MODULE HAS
LANDED ON MOON

ROBERT HEINLEIN
SCIENCE FICTION WRITER

DISNEYLAND

DON WEBSTER
CBS NEWS
SOUTH VIETNAM

The New York Times

The New York Times
MEN LAND ON MOON

MAN
ON THE
MOON

Lunar Day: Broadcasting, the Press, and Apollo 11

"Of all humankind's achievements in the twentieth century—and all of our gargantuan peccadilloes as well, for that matter—the one event that will dominate the history books a half a millennium from now will be our escape from our earthly environment and landing on the Moon."[1] —Walter Cronkite, CBS News

THE APOLLO 11 lunar landing was a television story. More than 53 million homes with television—94% of all American homes—witnessed some portion of the three major networks' coverage of the Apollo 11 mission. During daylight hours, viewers increased by 77%; during primetime, 42% more viewers watched the coverage than would have tuned in to regular network programming. Hundreds of millions more watched around the world. In the era before home video recording, commercial DVDs, or online videos, the only way to experience the coverage was to watch it live. Everyone was aware that television images were fleeting, and there was no assurance they would be preserved. (In fact, in the years since, many broadcasts from that decade have been lost.)

CBS News Anchor Walter Cronkite, a noted space enthusiast, who covered twenty-one manned space flights in his career, including all leading up to Apollo 11, had started covering space during the forerunner unmanned programs. Indeed, as a United Press print reporter during the last days of World War II, he witnessed plumes of smoke over the Netherlands as German V-2 rockets traveled from their launch site in Wassenaar on the way to London. Later, in the swamps of Florida, he covered for CBS the fledgling U.S. space program, then run by the U.S. Army. At the time, reporters were merely tolerated, forced to guess when launches would occur by divining information from clues such as the presence of searchlights or when high-ranking military officials checked in at local hotels.

The Eisenhower White House's bold decision to place the American space program under civilian control and institute an open-press policy was not lost on Cronkite. "When it was decided that the country should plan for manned flight as well as the perfection of ballistic missiles, it was also wisely conceded that such an expensive program was going to need public support and that this would be hard to get in an atmosphere of secrecy," he wrote in his memoirs. "Hence the National Aeronautics and Space Administration was born and the program, in most phases, opened to the press."[2]

The U.S. program's open policy—a savvy piece of public affairs marketing during the height of the Cold War—and the decision to divert massive capital toward a goal without a clearly defined strategic outcome, prompted CBS News national correspondent Eric Sevareid to say on the eve of Apollo 11: "The great debate about America in space is an exercise in freedom, the freedom of choice. How shall a people use its excess energies and resources?"[3]

All three major U.S. television networks poured immense resources into their live Apollo 11 coverage, but the effort by CBS News, as overseen by Executive Producer Robert J. Wussler, was acknowledged by contemporary television critics as the standard of excellence. Heralded by a dramatic animated graphic, which climaxed with an image of the Earth and Moon in alignment—clearly inspired by the opening of Stanley Kubrick's film *2001: A Space Odyssey*—CBS called its live coverage "Man on the Moon: The Epic Journey of Apollo 11."

Throughout the 1960s, the legacy of the Edward R. Murrow years was still held as a benchmark by most employees at CBS News, and many of the original contingent of "Murrow's Boys"—Eric Sevareid, Charles Collingwood, Winston Burdett, and Richard C. Hottelet—appeared on nearly every newscast. Publicly, CBS promoted its commitment to journalistic excellence and garnered much attention with the first expanded half-hour evening newscast in 1963. Nevertheless, not everyone within the CBS family was enthusiastic about extensive coverage of the manned space program. Network affiliates, in particular, resisted lengthy preemptions of lucrative program-

"MEN ON THE MOON: THE EPIC JOURNEY OF APOLLO 11." The CBS News coverage of Apollo 11, from liftoff on July 16 to splashdown on July 24, was an epic effort in its own right, marked by meticulous planning and extraordinary thoroughness. During and around the Moon landing, the network provided thirty-two hours of continuous coverage. Scenes that could not be photographed from space, such as the docking of the lunar module to the command module, were shown in simulation. Beside veteran reporters Roger Mudd, Mike Wallace, Dan Rather, and commentator Eric Sevareid, experts were gathered from many disciplines and included representatives from the contractor companies, scientists, science fiction writers, and astronauts. Retired Apollo 7 commander Wally Schirra joined Cronkite at the news desk as Special Analyst. For the CBS staff reporters, it was all hands on deck, as they covered public reaction to the events around the country and the world.

1. Walter Cronkite, *A Reporter's Life*. New York: Alfred A. Knopf, 1996.
2. Ibid.
3. CBS News and CBS Television Network. 10:56:20 PM EDT, 7/20/69: *The Historic Conquest of the Moon as Reported to the American People*. New York: Columbia Broadcasting System, 1970.

CBS monitors in Central Park, New York City, where tens of thousands watched the first Moon landing live.

ming for live news reports. Unlike Cronkite, CBS News President Richard S. Salant, was vocally ambivalent about the network's commitment to coverage of manned space flight and, in 1966, questioned the wisdom of preempting hours of regularly scheduled programming for what he described as "marking time, making brownie points, backing and filling when the mission is routine."[4]

It was clear to everyone in July 1969 that the flight of Apollo 11 was going to be anything but routine. And in anticipation, CBS News spent months preparing for the epic moment, which came as part of the network's continuous thirty-two-hour of coverage of *Eagle's* landing, the EVA (moonwalk) and departure from the Moon. Dubbed "Lunar Day" by CBS, the broadcast called upon a film library of 140 separate prerecorded segments that provided background information about the historic mission. These ranged from mini-biographies of the crew members and an exclusive interview with former President Johnson, to an essay on the Moon's place in human mythology and its depiction in science fiction films.

At the anchor desk, Cronkite was aided by commentary and reflection from Sevareid, recently retired Apollo 7 commander Wally Schirra, and science fiction author and visionary Arthur C. Clarke. Around the country, domestic correspondents filed remote reports from Houston, Disneyland's Tomorrowland, Washington's Smithsonian Institution, a lunar landing training site in Flagstaff, Arizona, North American Rockwell's plant in Downey, California, and Grumman in Bethpage, New York. A number of CBS News's veteran correspondents, including Mike Wallace, Winston Burdett. Marvin Kalb, Peter Kalischer, Morley Safer, Daniel Schorr, and Bob Simon covered world reaction from London, Rome, Paris, Amsterdam, Manila, Tokyo, Saigon, Belgrade, Bucharest, Mexico City, Montreal, Lima, and Buenos Aires as well as the International Arrivals Building at New York's Kennedy Airport. In all, more than 1,000 people were involved in the coverage.

In New York City, a set was constructed in Studio 41, CBS's largest sound stage, which normally was home to two soap operas. Cronkite's anchor desk was elevated twenty-four feet above the floor and was set against an artist's conception of the Milky Way galaxy with a six-foot diameter Moon globe and a similarly sized Earth globe. Sixteen live television cameras were deployed throughout the studio to cover the anchor desk; a status desk, where David Schoumacher would give up-to-the-minute reports on the mission's progress; and an interview area. When called for, live cameras would provide insert shots of spacecraft models, and one camera was continually fixed on a clock recording the mission's elapsed time, which was superimposed over the live broadcast whenever relevant.

The challenge of thirty-two hours of continuous coverage utilizing 142 television cameras was daunting. (The previous record for CBS News had been fifty-eight cameras for the 1968 Republican National Convention in Miami Beach, Florida.) CBS News Vice President Gordon Manning advised his team prior to the broadcast: "Let's follow that old journalistic maxim: 'Plan for the things you can plan for, so you are ready for the things you can't plan for.'"

Cronkite was normally a man with no trouble finding words to describe historic events. In what many consider the central moment of his career, Cronkite personified composed professional competence as he had to announce live to the nation the death of President Kennedy in November 1963. In July 1969, seemingly under much less immediate pressure than on that tragic Friday in 1963, Cronkite was temporarily, and astonishingly, at a loss for words during *Eagle*'s landing:[5]

CAPCOM [*capsule communicator, Charlie Duke*]: 60 seconds.

EAGLE: Lights on. Down 2½. Forward. Forward. Good. 40 feet, down 2½. Picking up some dust. 30 feet, 2½ down. Faint shadow. 4 forward. 4 forward, drifting to the right a little. 6 down a half.

4. Cited in Alfred Robert Hogan, *Televising the Space Age: A Descriptive Chronology of CBS News Special Coverage of Space Exploration from 1957 to 2003*. Masters Thesis, College of Journalism, University of Maryland, College Park, 2005.
5. Transcript of CBS News broadcast, July 20, 1969.

CRONKITE: Boy, what a day.

CAPCOM: 30 seconds.

EAGLE: Contact light. Okay, engine stop. ACA out of detent. Modes control both auto, descent engine command override, off . . .

SCHIRRA: We're home!

CRONKITE: Man on the Moon!

CAPCOM: We copy you down, Eagle.

EAGLE: (*Armstrong*) Houston.

SCHIRRA: Oh, Jeeze!

EAGLE: Tranquility base here. The *Eagle* has landed.

CAPCOM: Roger, Tranquility, we copy you on the ground. You've got a bunch of guys about to turn blue. We're breathing again. Thanks a lot.

CRONKITE: Oh, boy!

TRANQUILITY: Thank you.

CAPCOM: You're looking good here.

CRONKITE: Whew! Boy!

EAGLE: We're going to be busy for a minute.

CRONKITE: Wally, say something, I'm speechless.

SCHIRRA: I'm just trying to hold on to my breath. That is really something . . .

EAGLE: Very smooth touchdown.

SCHIRRA: Kind of nice to be aboard this one, isn't it?

CRONKITE: You know we've been wondering what these guys Armstrong and Aldrin would say when they set foot on the Moon, which comes a little bit later now. Just to hear them do it we are left absolutely dry-mouthed and speechless.

Later, while Cronkite and Schirra sat at the anchor desk, Armstrong made his decent down the ladder to the lunar surface. Cronkite was just as enthusiastic as during the landing, but this time had no trouble finding his words. In fact, there was so much studio commentary and excitement, Armstrong's first words on the lunar surface were lost for a few seconds.

ALDRIN: Roger, TV circuit breaker's in. Receive loud and clear.

CAPCOM: Man, we're getting a picture on the TV.

CRONKITE: There it is.

ALDRIN: Oh, you got a good picture. Huh?

CAPCOM: There's a great deal of contrast in it, and currently it's upside-down on our monitor, but we can make out a fair amount of detail.

CRONKITE: Two hundred million people are turning up on their heads at those words. They're supposed to turn that picture over for us.

ALDRIN: Okay, will you verify the position, the opening I ought to have on the camera.

CRONKITE: They've turned it over now.

SCHIRRA: There's that foot coming down now.

CRONKITE: There he is. There's a foot coming down the steps.

CAPCOM: Okay, Neil, we can see you coming down the ladder now.

ARMSTRONG: Okay, I just checked, getting back up to that first step, Buzz, it's not even collapsed too far, but it's adequate to get back up.

CAPCOM: Roger, we copy.

ARMSTRONG: It takes a pretty good little jump.

CRONKITE: So there is a foot on the Moon! Stepping down on the Moon! If he's testing that first step he must be steeping down on the Moon at this point.

CAPCOM: Buzz, this is Houston. F 2 1⁄160 second for shadow photography on the sequence camera.

ALDRIN: Okay.

ARMSTRONG: I'm at the foot of the ladder. The LM foot pads are only depressed in the surface about one or two inches. Although the surface appears to be very, very fine grained, as you get close to it. It's almost like a powder. Now and then, it's very fine.

CRONKITE: Boy! Look at those pictures! Wow! It's a little shadowy, but he said he expected that in the shadow of the lunar module. Armstrong is on the Moon!

ARMSTRONG: I'm going to step off the LM now.

CRONKITE: Neil Armstrong, a thirty-eight-year-old American standing on the surface of the Moon! On this July twentieth, nineteen hundred and sixty-nine.

ARMSTRONG That's one small step for man. One giant leap for mankind.

SCHIRRA: I think that was Neil's quote. I didn't understand it.

CRONKITE: Yes, "One small step for man," but I didn't get the second phrase. If some one of our monitors here at Space Headquarters was able to hear that we would like to know what it was.

REEL III SIMULATIONS. *Most of the Apollo story unfolded far removed from television cameras. To show what was happening during the live broadcasts, the three television networks routinely relied on spacecraft models, live action simulation, and animated depictions of events in space. In preparation for the Apollo flights, CBS commissioned highly accomplished animated footage from Reel III Animation, a firm overseen by Richard E. Spies, a veteran aerospace illustrator. Reel III's rendered depictions showed rocket firings in space, lunar descents, and fiery reentries. Partnering with Spies on these sequences at the three-man firm was a former Boeing technical illustrator named Ralph McQuarrie. Seven years later, McQuarrie was sought out by filmmaker George Lucas to envision the director's concept for his newest project, a trilogy of science fiction adventure films. It was McQuarrie's pre-production paintings for Lucas that propelled and inspired the look of what became* Star Wars. *McQuarrie is now widely credited for creating the appearance of Darth Vader, Chewbacca, R2-D2 and C-3PO and many of the film's sets.*

NBC's Huntley and Brinkley Cover Apollo

Throughout most of the 1960s, Walter Cronkite's *CBS Evening News* placed second in the national television ratings to NBC's *Huntley-Brinkley Report*, co-anchored by Chet Huntley, in New York, and David Brinkley, in Washington. A favorite of journalists and winner of eight Emmy Awards, the *Huntley-Brinkley Report* was the place a majority of American households went for a dinner-hour news summary during the formative era of television news. While Huntley and Brinkley excelled at covering political conventions, the machinations of Washington, the civil rights movement, and other moments during the '60s, the news duo seldom won plaudits for their broadcasting about the manned space program.

Television news historian Barbara Matusow described Huntley and Brinkley as "thoroughly bored" by the space story[1] and quotes veteran NBC News producer Jim Kitchell as less than impressed with his network's star anchors. "Brinkley felt it was too technical a subject for him. [NBC correspondent Frank] McGee, of course, did his homework, but I couldn't trust Huntley and Brinkley. I would provide them with all the research and information, and they wouldn't read it. They'd read the summaries in the press kits."[2]

Off screen, Huntley and Brinkley made few friends at Cape Kennedy or in Houston. In a 1990 conversation with historian Don Carleton, Walter Cronkite recalled that, when Brinkley visited Cape Kennedy, his detached and acerbic demeanor didn't win him any favors.[3] In his 1995 memoir, Brinkley disparages an unimaginative NASA bureaucrat and recalls his own boredom covering an Apollo mission in Houston. He describes wryly an evening motel pool party in Houston, during which Huntley rolled a piano into the swimming pool: "There was some drinking."[4] Huntley later confessed that covering the astronauts was "an exercise in boredom. The networks all got trapped. Most astronauts are as dull as hell, nice guys, mechanics. The only ones who had a mind of their own didn't last long."[5] Brinkley admitted that he found the space program difficult to cover on television, because he didn't consider it a visual story. "Easy to write, easy to talk about, but hard to show." He believed NBC was much too reliant upon NASA-controlled images to create the narrative.[6]

In contrast, Walter Cronkite's infectious enthusiasm for the space program won him new viewers. As Matusow noted, "by not establishing themselves as experts on space, Huntley and Brinkley abdicated the screen at moments when millions of additional people were watching. . . . Space was one of the few positive stories of the tumultuous sixties, and Cronkite's close identification with it helped further boost his image as the nation's patron saint and guardian."[7] By 1970 Cronkite's evening newscast had overtaken NBC's lead, a position it maintained for the next decade.

1 Barbara Matusow, *The Evening Stars: The Making of the Network News Anchor*. Boston: Houghton Mifflin, 1983, pp. 127–128.
2. Matusow, ibid.
3. Walter Cronkite and Don Carleton, *Conversations with Cronkite*. Austin: University of Texas Press, 2010, p. 230.
4. David Brinkley, *David Brinkley: A Memoir*. New York: Knopf, 1995, p. 197.
5. Huntley quoted in a 1970 *Life* interview cited in Lyle Johnston, *"Good Night, Chet": A Biography of Chet Huntley*. Jefferson, NC: McFarland & Company Inc., 2003, p. 84.
6. Brinkley, op. cit., p. 196.
7. Matusow, op. cit., pp. 127–128.

ARMSTRONG: The surface is fine and powdery. I can–I can pick it up loosely, with my toe. It does adhere in fine layers like powdered charcoal to the sole and sides of my boots.

CRONKITE: That was "One small step for man, one giant leap for mankind."

ARMSTRONG: I only go in a small fraction of an inch. Maybe an eighth of an inch, but I can see the footprints of my boots and the treads in the fine sandy particles.

CAPCOM: Neil, this is Houston. We're copying.

SCHIRRA: Oh, thank you television for letting us watch this one!

CRONKITE: Isn't this something! 240,000 miles out there on the Moon and we're seeing this.

The Nielsen ratings of the Apollo 11 coverage from launch to splashdown indicate that during the thirty-nine hours and twenty-five minutes when all three television networks were covering the flight, CBS commanded a 24% larger audience than NBC and 179% larger than ABC. When the astronauts walked on the Moon, more Americans watched on CBS than on the other two networks combined.[6]

Despite the excellent Nielsen ratings and the fact that all the sponsorships had been sold—one-third to Western Electric, one-third to International Paper Company, and one-sixth each to Kellogg and General Foods—CBS News's Apollo 11 coverage resulted in a $2.5 million loss for the network as a result of the extensive production costs. However, by winning the ratings war and earning the critical accolades of the critics with one of the most-watched television events of all time, CBS could once again claim the "Tiffany Network" nickname, a familiar brand recognized for quality and accuracy.

To promote their brand and commemorate their epic Apollo coverage, CBS News commissioned an illustrated, custom coffee-table book, *CBS News 10:56:20PM 7/20/69*. Released in 1970, the 169-page hardcover keepsake was given as a corporate gift to NASA leaders, members of Congress, foreign heads of state, college libraries, favored clients, VIPs, advertisers, and affiliate executives.[7] The book still provides an excellent overview of the historic conquest of the Moon as reported to the American people by CBS News. The images reproduced in the book are all shot from television screens and much of the text is from transcripts of on-air coverage by Walter Cronkite, Roger Mudd, Mike Wallace, Dan Rather, and special correspondent Wally Schirra.

Included in many copies of the book was the business card of one of several top CBS executives, such as Frank Stanton, President of CBS, Robert D Wood, President of CBS Television Network, or Richard W. Jencks, President of CBS/Broadcast Group, and each included a note on personalized letterhead. The one from Stanton reads: "This limited edition is the record of how CBS News met the most demanding challenge yet put to electronic journalism. We publish it with pride—in the accomplishment of Apollo 11 that we share with all Americans and the pride of achievement that we at CBS count among the great chapters in our history."

Designed by Louis "Lou" Dorfsman, the noted graphic designer who oversaw almost every aspect of the advertising and corporate identity for CBS during his forty years with the network, the book also celebrates CBS's iconic style. Dorfsman's dust jacket features no front title, just an embossed map of the lunar surface upon which is printed a small inset panel image of Aldrin and the American flag. The book recaptures a moment of high emotion and celebration for all Americans, while also reminding its select recipients, "We did a great job covering Apollo 11. You can trust us with your business going forward."

The commemorative volume CBS News 10:56:20PM 7/20/69, *published in 1970, featured a transcript of highlights from the Apollo 11 flight and a color center section of parallel scenes from the video coverage, all arranged to give the reader the impression of minute-by-minute coverage of the events. The scenes shown above are from the last minutes of July 20 and the first hours of July 21. In addition to the scenes broadcast live from the moon, we see President Nixon speaking with the astronauts, a view of Neal Armstrong's mother, Mrs. Stephen Armstrong, watching the events, Apollo 8 astronaut William Anders, CBS commentator Eric Sevareid, and Walter Cronkite at the news desk with astronaut Wally Schirra, who provided commentary throughout the broadcast.*

6. *CBS News 10:56:20 PM EDT, 7/20/69*, op. cit.
7. Alfred Robert Hogan, op. cit.

A 1970 NASA photo of the media stations that were set up for the Apollo 13 mission in the lobby of Houston's Manned Spacecraft Center's Building 1. A NASA contractor, A-V Corporation, is filming the fast-paced activity of the newsroom for the official Apollo 13 mission documentary. Standing in the center is NASA Public Affairs Office staffer Don Peterson. Seated at the table is Reuters reporter Mary Bubb, the first female reporter to cover the space program.

OPPOSITE: A 1971 photo, taken during the Apollo 14 mission, also at the MSC in Houston, in the lobby outside the auditorium used for formal press conferences.

Naturally, on the occasion of the book, CBS News President Richard Salant had only praise:

> For my colleagues at CBS News, and for myself, covering "Man on the Moon: The Epic Journey of Apollo 11" ranks as the single most satisfying effort in our collective experience as journalists. All too often we are forced to report man's shortcomings. In this instance, from the moment of blast-off to the moment of splashdown we were continually conscious of being involved in one of the great triumphs of the human spirit.

Apollo 11 was also a turning point in network coverage of manned spaceflight and at CBS News. In the months that followed it, the network news divisions faced corporate pressure to cut back live coverage of the later Apollo missions. Whether this was the cause of viewer apathy or the result of it has been much debated.

Only three years after Apollo 11, reflecting on "the single most satisfying effort in our collective experience as journalists," Salant expressed a very different opinion about CBS News's coverage of Apollo in an internal memo to colleagues:

> Let me put it bluntly: I do not think that Apollos are any longer prime news and nobody has told me anything about this flight [Apollo 17]—except that it is the last one and since it is at night, it will be pretty visible and spectacular—which makes it more than routine. Sorry, but in the absence of further information or further persuasion, my initial instinct is to interrupt [the prime-time drama series] *Medical Center* five minutes before the scheduled launch and go back to *Medical Center* as soon as the Apollo 17 is no longer visible to camera. Further, I would like somebody to explain to me why live splash down is worth the couple of hundred thousand dollars it would cost."[8]

How the Other Outlets Did It: Radio and the Newspapers

Images of the Mission Control room at Houston's Manned Spacecraft Center were a familiar sight to any casual viewer tuning in to television coverage of flights during the Apollo era. Nearby was another room buzzing with activity, seldom seen in television coverage. There, credentialed journalists noisily pounded out stories on manual typewriters as the clatter and pop of keys striking paper and ringing carriage return bells sounded throughout the room. Not far away, squawk boxes broadcast Mission Control's air-to-ground audio communications, and reporters dictated dispatches into telephones to their home offices.

8. Alfred Robert Hogan, op. cit.

As 21st-century communications technology evolves at an astonishing pace, recreating a detailed picture of news gathering efforts a few decades earlier tends to be lost to time. Ironically, the very technology that evolved in part as a result of the space program has blurred a contemporary view of the recent past. The remarkable feat of broadcasting live television images of humankind's first steps on another world—and the stunning quality of the later color television from the lunar surface—tends to suggest that reporting on the Apollo missions was a fairly easy news assignment. However, print and broadcast journalists covering Apollo from NASA's press rooms or from contractor-hosted press centers like the Joint Industry Press Center, in Houston, and the Apollo Contractors Information Center, at the Cape, were working with the technology of the day, frequently improvising to get the story out. When restricted by a very tight deadline, or when breaking news so required, many print reporters dictated their stories over the phone in a manner much like reporters depicted in Hollywood films of old.

Television and radio journalists had different technological concerns, usually involving the audio and video feeds that enabled them to report live from the Cape or from Mission Control in Houston. While many media outlets, notably the major television networks, spent large sums of money covering the Apollo program, the vast majority of reporters from fifty-seven nations covering Apollo 11 did so on tiny budgets, sometimes with very few resources.

Reporters such as Benjamin Wong from *El Sol de Mexico*, Jack Viviers representing *Die Burger* of Cape Town, South Africa, and Freda Gulick from WIS-TV of Columbia, South Carolina, often shared rental cars or slept on friends' sofas. And those were the registered media. Countless other publications and broadcast outlets reported from the Cape or Houston without official credentials.

Apollo 11 on a Shoestring

"This is KMHT in Marshall, Texas. We now go live to NASA's Manned Spacecraft Center and Wayne Harrison who is covering Apollo 11 . . ."

Wayne Harrison's KMHT radio introduction sounds the same as those heard on much larger stations, but the effort to pull it off on a minuscule budget is evidence of the passion some reporters like Harrison had for the Apollo program and the tenacity they showed in dispatching their stories.

Press scenes from the Manned Spacecraft Center in Houston. Clockwise from top left:

An MSC Public Affairs Office staffer answers questions from members of the foreign press during the Apollo 13 mission.

In 1970, typists at Western Union terminals send stories to newspapers around the world.

During missions, each shift change would be marked by a press conference intended to bring reporters up to date. Here, during the Apollo 15 lunar landing, Flight Director Gerry Griffin and PAO staffer Terry White hold a press briefing in Building 1 of the Manned Spacecraft Center in Houston.

Henry Simmons of Newsweek *(foreground) and John Riley of NASA Public Affairs monitor the news and activity in the Mission Control Center in Houston during critical moments of the Apollo 13 mission.*

An unidentified reporter intensely follows air-to-ground communication during the Apollo 13 mission.

Reginald Turnill of the BBC makes a live news report during Apollo 13, standing in front of the Apollo 9 command module.

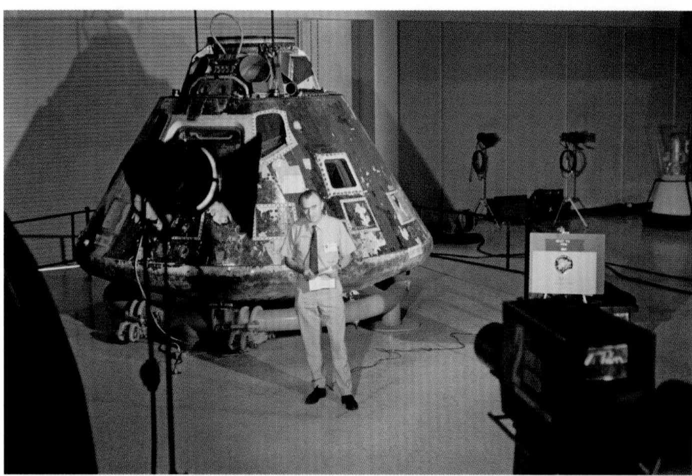

Harrison was an elementary school student when the first Mercury missions flew, and by the time he was a junior in high school he was already working part-time in commercial radio. Upon graduating from high school in 1967, he went to work for KMHT, a 1000-watt AM radio station in Marshall, Texas. "The person that I wanted to be most like at that time was Jules Bergman, the ABC News science reporter who extensively covered the space program. I started firing off letters to NASA asking for tapes and packets, whatever information they had, and told them that we were covering the space program."[9] KMHT radio reached approximately 30,000 people, and together with KLUE and KHER-FM, both in Longview, Texas, made up the "Big K Network."

The young reporter wanted desperately to cover the Apollo 11 lunar landing, which Harrison believed was the biggest story that was ever going to come his way. While he had covered Apollo 8, the first manned lunar mission the previous year, the owner of the Big K Network would not approve the budget necessary to send him to Houston (220 miles away) for Apollo 11. NASA's news audio, which was delivered to KMHT, was all he had to work from. But Harrison wasn't satisfied.

> I arranged my vacation so that I would be down in Houston. We did what's called a "trade-out," a barter arrangement with a hotel in Galveston, thirty miles away. In exchange for free radio advertising, I got a room. So the only thing the station owner was paying for was long-distance phone calls. Then the station could say: "We have our own reporter on the scene." For me, it was an experience of a lifetime.[10]

During Apollo 8, Harrison attended the press conferences and collected press releases from NASA and the contractors. NASA provided communal tables where radio reporters could work and provided a connection to an audio "mult box," a broadcast splitter carrying NASA's live audio feed that reporters listened to with either a headphone or a speaker connection. Harrison wrote his reports on his own portable typewriter and then went to a pay phone in the hallway to file his live reports on the air. But there was something missing from his broadcasts—live audio of NASA mission controllers as they communicated with the astronauts.

The major radio broadcasting outlets—and a number of medium sized outlets as well—had first-class setups. Many erected small audio booths that featured the live NASA feed, dedicated telephone lines, a broadcast microphone connected directly to their network, and a comfortable place to work. "They were all wired up to go, and they had a budget," Harrison recalled. "I had no budget because I was there on a freebie, so I had to use the pay phone. My problem was: I'm in the hallway away from my desk. How can I hear what's going on?"

By the time of Apollo 11, Harrison figured out a way to provide a live broadcast that sounded as professional as those heard on the major outlets, but merely for the price of a collect call. He smuggled his own equipment into the press center, which included a long length of cable and a small battery-operated speaker with a volume control. He connected his speaker to the mult box and strung the line out to the pay phone in the hallway.

> I was afraid that the phone would get tied up, because I only called the station when something significant was happening. So a buddy of mine who worked for the telephone company lent me an "out-of-order" sign to hang on the phone. My other problem was that I was afraid NASA would get upset about me running this long cable down the hallway, so I wound it up and I hid it and the speaker behind a Coke machine in the press room. When I went to file a report, I grabbed my speaker from behind the Coke machine, ran the cable down to the phone, removed the "out-of-order" sign, and put the speaker next to the phone. Then I could call the radio station collect and hold the phone down to the speaker when I wanted to pick up the NASA audio to put it live over the air.[11]

The innovative spirit embodied in the more than 400,000 people behind the Apollo missions—from NASA employees and contractors to university professors and communications and telemetry specialists abroad—was reflected in the pluck of journalists like Harrison, who traveled to the Cape and to Houston, sometimes paying their own way. For many of Harrison's generation, the lunar missions were understood to be a once-in-a-lifetime opportunity; the chance to cover a story that would forever redefine humanity and its place in the universe. Despite the odds, young reporters like Wayne Harrison discovered how to cover the story, even when their news outlets couldn't afford it, and they did so in ways that would have made media veterans like Walter Cronkite proud.

9–11. Wayne Harrison, interview with the authors, January 19, 2012.

The ABCs of Being in Third Place

ABC News science editor Jules Bergman in 1973, in a promotional photograph prior to the first Skylab mission, which used hardware salvaged from cancelled Apollo Moon landings. Early in his coverage of the American space program, Bergman subjected himself to the same exercise regimen as the first astronauts and was filmed enduring 5 Gs in a NASA centrifuge, an experience he said he found "exhilarating."

Throughout most of the 1960s, ABC was the third-place network in both entertainment and news. Despite occasional hits like *Batman* and *Bewitched*, ABC was seldom the first choice for '60s viewers when turning on their sets. So when all three networks devoted hours of live coverage to the American space program, ABC's effort seldom garnered a sizable portion of the audience and, accordingly, the network's budget and commitment reflected this reality.

However, ABC did not lack for a consistent and familiar face during its live coverage. Serving as the network's science editor from 1961 until his death in 1987, Jules Bergman covered fifty-four manned space missions for the network, beginning with Yuri Gagarin's flight in April 1961 and concluding with the *Challenger* disaster in January 1986. Press materials seldom mentioned that Bergman's full name was Jules Verne Bergman, named after the French science fiction pioneer, who was reported to have been a tenth cousin.[1]

On air, Bergman could evoke the demeanor of a smart but irritating kid with a slight speech impediment, who sat in the front row of a high-school chemistry class and was never reluctant to call attention to an instructor's error. His reporting, though, was sober and well-informed, and before Walter Cronkite, Bergman, who was an experienced private pilot, subjected himself to the physical tests endured by the astronauts during training. As recounted in the *New York Times*, this even extended to his live television coverage: "While covering the Aurora 7 flight of M. Scott Carpenter—the second American to orbit the Earth—Mr. Bergman had doctors place him in a harness with sensors identical to those that monitored the astronauts. The instruments showed he was under as much stress during his twelve hours on the air as Mr. Carpenter was during the five hours of his three orbits."[2]

In contrast to Cronkite's optimism about the American space program, Bergman had a penchant for melodramatically underlining the physical dangers of space flight. During the initial hours of ABC's coverage following the oxygen tank explosion on Apollo 13, viewers who heard Bergman's coverage on ABC were unlikely to have thought it possible for the crew to survive.

For their coverage of Apollo 11, ABC News called upon Frank Reynolds to fill the anchor desk. Reynolds was, at that time, co-anchoring the network's evening newscast with veteran broadcaster (and Edward R. Murrow protégé) Howard K. Smith. Unlike Cronkite, Smith, and Bergman, Reynolds had never worked as a print journalist, having begun his career in local radio and television in Indiana. On screen, Reynolds projected a calm, affable, keen intelligence that served as a welcome counterpoint to Bergman's humorless intensity.

At ABC News there was no prohibition about on air anchors verbally mentioning an advertiser's name or physically placing the sponsor's logo on the anchor desk during their Apollo coverage. (CBS News had a policy that clearly separated advertising from their news content.) So it was not uncommon to hear Reynolds talking about the latest update from Houston while an orange sign for Tang or a rectangular Philco logo appeared strategically positioned on the set.[3]

Walter J. Pfister, ABC News Special Events Producer, and ABC News anchor Frank Reynolds prior to the network's live coverage of the Apollo 11 mission. Pfister was later to work with Reynolds on ABC News's special Watergate coverage, in 1974.

1 *Encyclopedia of Television News.* Westport, Conn.: Greenwood Press, 1998.
2. James Barron, *The New York Times*, February 13, 1987, "Jules Bergman, 57, Science Editor of ABC News for 25 Years, Dies."
3. The CBS policy is reported in Alfred Robert Hogan, "Televising the Space Age," Master's Thesis, University of Maryland, 2005.

Walter Cronkite, reporting during Apollo 11 with former astronaut Wally Schirra, holds up a copy of the New York Daily News *article of the Moon landing, written by science writer Mark Bloom and published on the morning of July 21, 1969.*

MAN Lands on the MOON

Over at the *New York Daily News*, veteran journalist Mark Bloom was also reporting the news of the Apollo 11 landing and first moonwalk, but for the biggest circulation newspaper in the country; in 1969, the *Daily News* could boast of a circulation of more than two million on weekdays, and more than three million on Sundays. Bloom got his start covering the manned space program as a Reuters wire service reporter. Based in the New York bureau, he covered all ten Gemini missions. In late 1966, just as the Gemini program was coming to an end, Bloom signed on as the *Daily News*'s science writer. "The primary requisite was to know something about the space program," he recalled decades later. "I would be covering medicine and other science stories, but we were going to go to the Moon and the *Daily News* needed someone to cover Apollo. Reuters offered to keep me by saying if I stayed through Apollo, they would make me bureau chief in Saigon, which was a big story in those days, but I went to the *Daily News*."[12]

While Bloom had resources at his disposal that were the envy of many of his competitors, he understood that his assignment required him to do more than merely report on the biggest story of the century. He also had to convey the meaning of the moment to his millions of readers. His choice of a headline or a turn of phrase would have to capture the moment so effectively that it would eclipse the competition. It was reporting that would serve not only the story, but history as well.

During the Apollo missions, Bloom worked out of the print newsroom at Manned Spacecraft Center in Houston. Unlike the inexpensive makeshift supplies Harrison used to file his live radio reports, Bloom benefited from the support of his New York editors. He enjoyed a private workspace with his own squawk box speaker provided by NASA, his own dedicated telephone, in addition to the available television monitors throughout the room. "The first edition deadline for the *Daily News*—what they used to call the 'one star'—was at 5:00 P.M. But, during the evening, if things broke, you could insert new copy into the story, or you could file a new lead."[13] Bloom wrote at his workstation on an Olivetti Lettera 22 portable typewriter, and when he had a story, Western Union representatives were ready. "Western Union was a kind of an unsung hero of all these missions," he admitted. "They were there hovering over you. You would yell, 'Western!' and they were there. They'd grab the copy and file it to New York via Teletype. You'd write on the top 'NPR collect'—meaning 'night press rate collect'—or 'DPR collect'—'day press rate collect.'"[14] Because print reporters were handing over their copy on typewritten pages, the transmittal process required Western Union employees to re-key the stories into a Teletype machine which, in turn, printed out a hard copy at the newspaper's offices, where it was edited and set on their Linotype machines (some of which were fitted to accommodate direct Teletype punch-tape input).

> It went directly into the wire room of the *Daily News*, where there was a copy boy assigned to my copy. I kept a carbon, but I assumed that what I wrote was what I sent because Western Union was very good. The copy boy at the *Daily News* in New York would rip the copy from the Teletype machine and take it to the national desk for editing, and then to the copy desk, who would also go over it. The editors used a red pencil to mark up the copy directly on the Teletype printout, and then the marked-up copy would be sent down to the composing room by a pneumatic tube system, where it was put into hot lead with an old Linotype machine for printing in the newspaper.[15]

This was the process Bloom used to file the front-page story that appeared on newsstands the morning of July 21, 1969. "I remember my lead," Bloom remembered decades later. "It was 'Man Walked on the Moon Today.' It was cosmic."[16] ⊙

12–16. Mark Bloom, interview with the authors, January 13, 2012.

In A Different Orbit: The Astronauts As Celebrities

FOR A BRIEF PERIOD in the American saga, the astronaut was the man of the moment. No profession commanded as much awe and admiration. Widely regarded as the personification of all that was best in the country, the first astronauts were blanketed with the adulation usually accorded star quarterbacks, war heroes, and charismatic movie stars. Yet this was never part of NASA's agenda. In fact, there were concerted early efforts to avoid such celebrity. However, the men chosen to be the first Americans in space were raised in a culture that revered the stoic aviator, and many saw themselves as the latest members of that select spiritual brotherhood.

Celebrated in headlines, fiction, and film, the leather-jacketed pilot personified a new aristocracy during the first half of the twentieth century: a young adventurer whose courage and daring bridged continents and cultures. Prior to World War I, fledgling aviators and their early airplanes were frequently depicted on the covers of mass-circulation magazines and on advertising posters throughout Europe. Shortly after hostilities commenced in 1914, nationalistic publications enthralled the public with stirring tales about the most decorated pilots and their bravery in the skies, feats far removed from the mechanized horror in the trenches below. During the interwar years, adventure magazines and cinema screens often featured romantic depictions of the solitary fighter pilot, silk scarf flowing in the wind, engaging his fellow knights of the air in aerial combat in the skies over France, while overhead across America, former military pilots risked their lives by supporting themselves as barnstormers and flying the first airmail routes.

Into this post-war culture arose aviation's first superstar, Charles A. Lindbergh, the story of whose solo transatlantic journey was known to every schoolboy in America during the decades that followed. The media attention that transformed Lindbergh from a unknown pilot from the Midwest into the most famous living person on the planet set the model—and offered a cautionary warning—for the American press's treatment of the astronauts a quarter-century later.

The initial group of seven astronauts got a sudden glimpse of their new public role on April 9, 1959, when they were first introduced to the world press in a crowded room in the Dolley Madison House in Washington D.C. The moment when the Mercury Seven first faced a barrage of clicking cameras and fielded unexpected questions from the Washington press corps is memorably recaptured in a key scene in Tom Wolfe's *The Right Stuff*. The press's insatiable curiosity reflected the country's fascination and hunger for heroes and, in John Glenn, many journalists wondered if NASA had found the Lindbergh of the space age.

"NASA wasn't out to market or create heroes," recalled Gemini and Apollo veteran Gene Cernan, who was introduced to the public as member of the third group of astronauts in 1963. "The hero status was created by the nature of what we did, and by the public's response to what we were doing."[1] Yet, as Apollo 7 astronaut Walt Cunningham wrote a few years after his mission, "In the glory years of manned space flight, what the country kept forgetting was that we were people. The only thing we did not see ourselves as was heroes . . . that was something the public craved, and the media had created—not that we didn't enjoy the myth, we just never fully understood it."[2]

Initially portrayed as both warriors in the fight to dominate outer space and modern-day exemplars of the idealized traits of the American pioneering spirit, the first astronauts quickly realized that their public images were largely a creation thrust upon them by an adoring press. This public portrait was further honed and developed as a result of the exclusive *Life* magazine profiles orchestrated—and, indeed, molded—by NASA's Public Affairs Office. At the same time, NASA tried to avoid making any one astronaut overshadow his comrades. There

OPPOSITE: Apollo 15 astronauts Jim Irwin, Dave Scott, and Al Worden appear with Dick Cavett on his late-night ABC network talk show in August 1971, three weeks after their return from the Moon.

1. Gene Cernan, interview with authors, 2012.
2. Walter Cunningham, *The All-American Boys*, New York: Macmillian, 1977, p. 161.

LEFT: *John Glenn was inundated with fan mail after becoming the first American to orbit the Earth in February 1962. In this vintage letter from June 1963, Glenn apologizes for the lateness of his reply, explaining that he "received over 100,000 letters during the past year." Some astronauts tried to answer all of their fan mail; others called upon help from NASA Public Affairs to respond. Glenn received so much interesting and varied fan mail following his three-orbit flight that Life/World Book Science Services, as part of their exclusive contract with the Mercury astronauts for rights to their private stories, published a book of correspondence with comments from Glenn entitled,* Letters to John Glenn: P.S. I Listened To Your Heartbeat.

RIGHT: *The astronauts were often celebrated with ticker tape parades in major cities and hometowns following their return to Earth. Here, Mercury astronaut Gordon Cooper and his wife ride in the "Welcome Home" parade in Houston, Texas. Thousands filled the street to greet Cooper after his 22-orbit flight of 1963, at the time the record for the longest duration of an American spaceflight.*

NATIONAL AERONAUTICS AND SPACE ADMINISTRATION
MANNED SPACECRAFT CENTER
HOUSTON 1, TEXAS

IN REPLY REFER TO:

JUN 19 1963

Mr. George R. Sanders, Jr.
President, The Exchange Club
7395 S. W. 87th Avenue
Portland 23, Oregon

Dear Mr. Sanders:

Thank you for your letter requesting a picture for your Father. In accordance with your request, I am happy to send the enclosed picture which I hope is satisfactory. I certainly consider it an honor to have my picture displayed with the pictures of the dignitaries you mentioned.

I regret my delay in replying to your letter. However, I have received over 100,000 letters during the past year, and I am just now beginning to get caught up with my correspondence.

Best regards to you and your Father.

Sincerely,

John H. Glenn, Jr.
Lt Colonel, USMC
NASA Astronaut

BEER
Welcome
(HO1)HOUSTON, TEX. MAY 23--COOPER WELCOMED--Astronaut Gordon Cooper and Mrs. Cooper ride down Main Street in Houston, Tex., today as thousands greeted them. (AP Wirephoto) (dt51630stf) 1963

was a concerted effort to define them as members of a unified team of pilots and technicians. Indeed, Leo Braudy in his essential history of celebrity,[3] suggests that NASA's insistence on a heroic team of astronauts and technicians was directly informed by knowledge of the disastrous political fortunes of Lindbergh, who in his highly public role as a private citizen urged accommodation with Hitler's Germany on the eve of World War II.

As the space race progressed, and NASA executed a number of "space firsts" during the Gemini program, the names of the second and third group of astronauts became nearly as well known. As perplexing as it was for the astronauts, especially the newer members of the team, it was part of the job—an occupational hazard that would follow them the rest of their lives. "No one could have predicted the public fascination with astronauts from the first unveiling of the Mercury Seven in 1959, through project Apollo," wrote Roger D. Launius, a NASA Historian and Smithsonian Air and Space Museum Senior Curator for Human Spaceflight. "The astronauts as a celebrity and what that has meant in American life never dawned on anyone before. To the surprise and ultimate consternation of some NASA leaders, they immediately became national heroes and leading symbols of the fledgling space program."[4]

Wearing the mantle of instant celebrities—even if they hadn't yet flown in space—the astronauts were often confronted by situations in which well-wishers desired an autograph or a photo or a handshake. Mindful of their place as official representatives of NASA and the American space program, they invariably tried to accommodate the public. "You had to give them a little something—you had to try," said Cernan.

3. Leo Braudy, *The Frenzy of Renown: Fame and Its History*. New York: Oxford University Press, 1986, p. 25.
4. Roger D. Launius, "Heroes in a Vacuum: The Apollo Astronaut as Cultural Icon" 2005, a paper delivered at the 43rd AIAA Aerospace Sciences Meeting and Exhibit.

LEFT: Dr. Wernher von Braun was also accorded celebrity treatment as part of the American space program. Here Dr. von Braun greets an enthusiastic crowd seeking autographs at the Gulf South State Fair in Picayune, Mississippi, in October 1963.

RIGHT: Neil Armstrong besieged by autograph-seeking children at a public appearance in front of a display of NASA rockets and capsules at the 1964 New York World's Fair. The photographs held by the children were supplied by NASA Public Affairs personnel and are printed with a reproduction of Armstrong's signature. The children, much to Armstrong's obvious chagrin, are angling for the real thing.

"You couldn't turn your back and walk off. Everything depended on how you related to people." But sometimes the effort resulted in long lines and huge crowds, which easily destroyed the best of organized schedules of the Public Affairs Office. As the highest-profile figures in a space program being sold to the public, the astronauts could not afford to be seen in a negative light or get a reputation for rudeness. Therefore, the Public Affairs officer assigned to an astronaut was often required to play interference. He could put a limit on the available time an astronaut had to stand and mingle, informing the crowd that he had only five, ten, or fifteen minutes left—and thereby giving the hero a graceful exit. "They could hustle you in and out sometimes," Cernan recalled. "Sometimes you'd get people starting to form a line for autographs—even before you had flown a mission—like a rookie baseball card kind of thing. And so, the Public Affairs Office guy would hustle you through."[5]

During the early years of the manned space program, Shorty Powers and NASA's Public Affairs Office carefully crafted the images of the first Mercury astronauts, primarily through the *Life*/World Book contract. But by the time the Gemini program was underway during the mid-1960s, a change had occurred. The military control over immediate information and the astronauts' personal stories during the Mercury era disappeared with Powers's departure in 1963. Even though the *Life*/World Book contracts were still active, the increased influence of Julian Scheer and Paul Haney signaled a marked difference in Public Affairs management, and with it fewer restrictions on information given to the media. During Mercury and some early Gemini missions, mission audio transcripts were commonly edited and cleansed for language and clarity. In contrast, Apollo was provided to the world unedited and unpolished. As Cernan described it, "We didn't doctor up the movie, didn't edit anything out. What was said, was said."[6]

The instantaneous and uncensored release of information to the press placed severe limitations on anyone attempting to carefully craft a public image. And for the astronauts working under stressful conditions, in a new spacecraft and in situations that were both unscripted and often prone to unforeseen challenges, the Apollo program revealed how greatly things had changed as early as the first manned mission. During the Apollo 7 flight in October 1968, the world heard the cranky and cantankerous voice of veteran astronaut Wally Schirra commanding his last mission. He battled not only a head cold and a flight plan he thought was littered with too many extra experiments and frivolous activities, but he also battled openly with his colleagues down at Mission Control. The result was damage, not only to his image but also to his crew in the eyes of the media and NASA's management. "The heavy workload at the outset of the mission, combined with his discomfort, made Wally more irascible by the day," wrote Cunningham in his 1977 memoir.[7] "He didn't miss an opportunity to nail Mission Control to the wall. Donn and I were amazed at the patience

5. Gene Cernan, interview with the authors, February 24, 2012.
6. Ibid.
7. Walter Cunningham, op. cit., pp. 127–128.

Presented to members of The
Franklin Mint Collectors Society
AUGUST 28, 1971

***Top to bottom:** a Franklin Mint mini-coin, minted with metal flown to the moon on the Apollo 14 mission; one of the postal covers carried to the lunar surface aboard Apollo 15; a photo by the Apollo 15 astronauts of the Fallen Astronaut tribute on the lunar surface; the PPK bag of Gene Cernan, flown to the lunar surface on Apollo 17.*

Fallen Astronaut: The Astronauts' Image Comes Down To Earth

Throughout the course of the space program, the specter of commercialization occasionally caused negative PR flare-ups. For the astronauts, there were free dinners and drinks, access to special deals, or preferred seating[1]—the kinds of perks that no one begrudged them. But there was also the Houston businessmen who offered free housing, and the Mercury astronauts' ill-fated attempt to invest in a hotel near Cape Canaveral.[2] The *Life*/World Book contracts came under fire several times, but a Congressional inquiry ensued when the Franklin Mint issued mini-coins that included grains of silver from coins that had flown to the Moon on Apollo 14 in an astronaut's Personal Preference Kit (PPK). The kits were beta cloth pouches in which the astronauts could take along personal keepsakes and mementos for themselves, their family and friends, and for select ground personnel.[3] Alan Shepard, the commander of the mission through whom the mint had arranged the flight of the metal, publicly apologized, promising that nothing like this would happen again.[4] Occasional editorials decried the appearance of commercialization and of the astronauts "cashing in" on their fame, though none lasted beyond the news cycle.[5]

This changed, however, with a scandal that erupted following the Apollo 15 flight, in July 1971.[6] A German stamp dealer arranged through commander Dave Scott to take to the Moon one hundred numbered, first-day postal covers, to be signed by the three crew members. The dealer had promised each astronaut $7,000 in the form of savings accounts. Unbeknownst to officials, the crew had asked a Belgian sculptor to create a statuette in remembrance of the astronauts and cosmonauts who lost their lives in the furtherance of space exploration. The small aluminum sculpture, called "Fallen Astronaut," was left on the Moon next to the lunar rover at the end of EVA 3, along with a plaque bearing the names of the fallen.[7] The memorial was deposited while the cameras were turned off, as Dave Scott told Mission Control that he was doing cleanup work. The sculptor, Paul van Hoeydonck, had promised Scott that no copies would be made, though he subsequently advertised copies for public sale, an offer later withdrawn under pressure from NASA. A copy was made, however, for the National Air and Space Museum, and the original with its plaque were photographed by the crew of Apollo 15.

These actions quickly caught the attention of NASA, and sparked outrage in some corners of the press, and in letters sent to NASA and members of Congress. Critics of NASA and the space program pounced, using these events to further question government spending on a national space program. Further investigation by a Senate committee revealed that many astronauts were being paid $5 each for their autographs in a commercial venture with the same German dealer associated with Apollo 15; that practice, too, was immediately shut down. As a result of the intense scrutiny of commercialization by astronauts, NASA tightened the PPK rules, limiting personal items for subsequent flights and placing tighter restrictions on their disbursements of flown mementos.[8] The affair left a bitter taste in the mouths of the astronauts, NASA, Congress, and the public. Behind closed doors, the Senate committee berated NASA for not doing a better job of protecting the astronauts, while at the same time lauding the astronauts for their service and bravery. Perhaps a bit sanctimoniously, some Senators also questioned their own right of ownership to the state flags and mementos that were given to them by returning astronauts.[9] NASA formally and publicly reprimanded the Apollo 15 crew for not properly following protocol of listing all their souvenirs on a pre-flight manifest.[10] The finger-pointing and anger that resulted from the incidents compromised the image of the astronauts, far beyond any justification and to the ultimate detriment to the program.

The question of astronaut souvenirs continued to haunt the former Apollo astronauts well into the 21st century. In a number of tense and high-profile cases in 2011[11] and early 2012,[12] NASA attempted to retrieve the personal artifacts and mementos from retired astronauts. The issue of astronaut ownership of these items, and their ability to dispose of them as they saw fit, was finally clarified by the passage of House Resolution 4158, a bill to confer ownership rights of all mementos (excluding moon rocks) to the Mercury, Gemini, and Apollo era astronauts. It was passed by the House and Senate in September 2012, forty years after the original Apollo 15 investigations, and just weeks after the death of Neil Armstrong.[13]

1. Robert Kibbe, "Astronauts and Corvettes," Corvette Online, February 3, 2011.
2. Nicholas Chriss, "Should An Astronaut Trade on His Name?" LA Times News Service, September 23, 1972.
3. "No More Unscheduled Freight Will Ride Aboard Spaceships," Associated Press, March 1, 1962.
4. Dan Thomasson, "Moon To Sail Along Without Silver: Shepard Orders Medallions Dropped From Apollo Missions." Scripps-Howard News Service, September 2, 1971.
5. "Astronauts Give SEC Depositions: Offered Stock In Company," Associated Press, July 14, 1971.
6. Art Buchwald, "Shady Moon Deal Hinted," syndicated column, September 24, 1972.
7. Nicholas Chriss, op. cit.
8. "Space Souvenirs to Be Banned," United Press International, September 17, 1972.
9. U.S. Senate, Committee on Aeronautical and Space Sciences. Executive Session, Commercialization of Items Carried By Astronauts, August 3, 1972.
10. Thomas O'Toole, "Astronauts Drummed Out For Money-Maker Plan," *The Washington Post*, August 1, 1972.
11. Lee Speigel, "Edgar Mitchell, Former Astronaut, Sued by NASA for Trying to Sell Apollo Camera," July 1, 2011.
12. "NASA Questions Astronaut's Right to Sell Apollo 13 Memorabilia." Associated Press Wire Story. 8 January 2012.
13. Pete Kasperowicz, "Obama Signs Bill Giving Apollo Astronauts Ownership of Space Artifacts," *The Hill*, September 26, 2012.

of those in the control center with some of the outbursts that came their way. On the ground, they were well aware that every word of the air-to-ground communications was being fed directly to the press center, a fact of which we had not been informed. So Wally's bad temper was making big news back home." By piercing the veil on the earlier image of the astronaut as calm, cool, fun-loving adventurer, Schirra made the mistake of reminding everyone that he wasn't superhuman.

Even when the astronauts were not performing their job before the eyes and ears of world's media, they were still in the limelight. "As astronauts, we were usually the biggest dudes in the crowd," wrote Cunningham. "And frequently the only ones who couldn't really pay their own way." There were constant invitations wherever they traveled from businessmen, politicians, entertainers, and ordinary citizens—to attend parties, open houses, black-tie affairs, and the like. While the astronauts could not possibly accept all the invitations, they certainly accepted a good number of them. "In a convoluted way, NASA encouraged our socializing. We were mixing with the community and selling the program." But the invitations usually had a catch—a favor, or the opportunity to use the astronaut's celebrity to the benefit of the evening's benefactor. Sometimes it wasn't overt, but often it was, especially in the epicenters of Houston and the Cape. "There was the time we and the Schirras were attending a concert at the Houston Music Theater—on passes," Cunningham recalled. "We were introduced just before the show started and as we sat down, Wally smiled and whispered, 'We just paid for our tickets.'"[8]

In retrospect, it may seem amazing that access to the astronauts—among the biggest celebrities of the '60s and early '70s—reflected a bifurcated world: there was unfettered access down at the Cape or in Houston at social events, on the street or in a bar, while, at the same time, NASA Public Affairs and the Astronaut Office allowed only extremely limited access for official interview requests, especially in conjunction with a mission. "Access to the astronauts before a mission was extremely restricted due to their preparations," veteran NASA Public Affairs officer Doug Ward recalled. "At that time, media interactions were usually during press conferences at the Manned Spacecraft Center and the Cape. Only a very select few might get a one-on-one interview."[9]

Veteran reporters were cognizant of the intense hours of

Blue Moon

The live television camera was off, but the microphones were on during the most frightening moment of the Apollo 10 mission. As astronauts Stafford and Cernan flew the lunar module thirty miles above the Moon's surface, a mistakenly thrown switch suddenly sent the vehicle into a series of wild gyrations. It took the astronauts approximately thirty seconds to bring the spacecraft under control. At the height of the alarm, as the LM pitched forward and the sight of the lunar horizon passed repeatedly through the window, Gene Cernan was heard to utter, "Son of a bitch!" Cernan's exclamation was broadcast live around the world. Most of the press treated this linguistic space "first" as a non-event on the day it occurred. Yet when the crew returned, Cernan made a public apology. This was prompted in part by a public rebuke to NASA from the president of Miami Bible College, who complained that the astronauts had used "the language of the street" and should repent their "profanity, vulgarity and blasphemy."

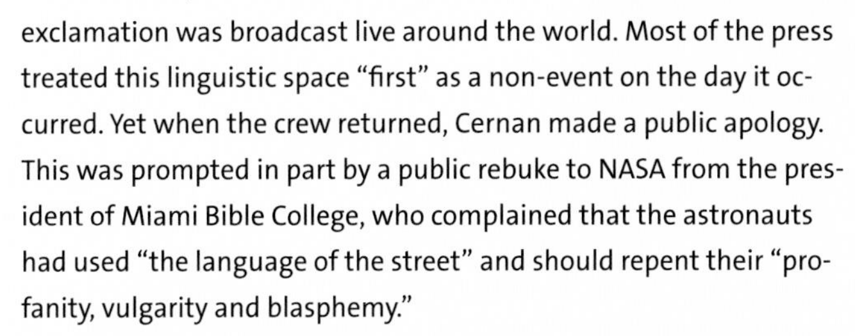

"I said it, and the whole world heard it," Cernan admitted. But unlike Schirra on Apollo 7, he addressed the issue in a post-flight press conference: "That was totally unrehearsed. What you heard was three men doing a job—sorry if we offended." He also addressed the matter with management, and issued a more formal apology. "I had to apologize, and I said, 'Can I apologize in my own way?' and they accepted it. I just said: 'To those I offended, I'm sorry; to those of you who understand, thank you.'" By addressing the issue, the impact wasn't as stinging as it was with Schirra. "You just have to let people know you are real," explained Cernan. "We were not any different. We put on our pants one leg at a time, just like everybody else."

The incident soon gained further notoriety, as it was recycled into late-night talk show monologues. Be that as it may, Cernan's apology effectively killed the story before it gained further traction, and NASA's Public Affairs Office quickly backed up Cernan with a reasoned response reported in *Time* magazine: "Those are human beings up there, and they acted like human beings. That's all, no more and no less."

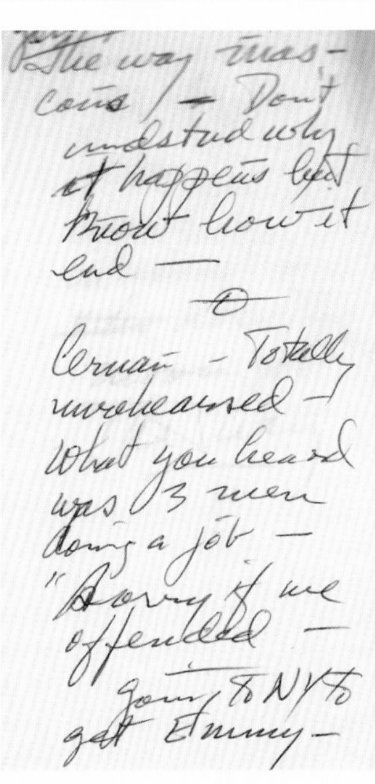

The way mascons / — Don't understand why it happens but know how it end —

Cernan - Totally unrehearsed — What you heard was 3 men doing a job — "Sorry if we offended" —

Going to NY to get Emmy —

TOP: *The Apollo 10 crew, left to right, Gene Cernan, lunar module pilot; Tom Stafford, commander; and John Young, command module pilot.*

ABOVE: *Paul Haney's notebook from Cernan's press conference.*

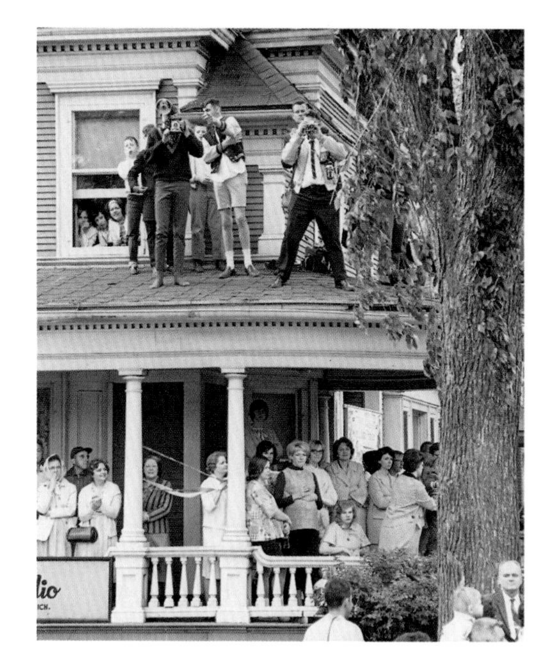

ABOVE: *Eager fans at a Detroit, Michigan, parade for Gemini astronauts Ed White and Jim McDivitt in 1965 climb rooftops to get a better view. (Photo: Detroit Free Press).*

RIGHT: *Well-wishers greet the crew of Apollo 11 in August 1969 at Chicago's Civic Center Plaza (now known as Daley Center Plaza) after a tickertape parade down LaSalle Street.*

training and preparation demanded of the astronauts in the weeks leading up to a mission, and they wouldn't even bother requesting an interview. But for the hundreds—sometimes thousands—of reporters on a first trip to cover a mission, there was often bitter disappointment. Ward remembers that the Public Affairs Office would always have to turn them down. "These guys would get a stricken look and say, 'Well I sold this to my editor on the assumption that I was going to get an astronaut interview.'" It just wasn't possible. But that didn't stop some of the reporters. "I remember, before Apollo 11 we had a guy come in from South America. He was just desperate and when he found out he couldn't get access to the astronauts, he staked out their homes. We got a report from the local sheriff that he'd been turned in. He climbed up a tree adjacent to one of the astronaut's houses and was trying to take pictures through the window. And John McLeish from our office had to speak to the sheriff and get the guy out of jail. Although there were rare occasions like that, the regulars who covered this thing as a full-time responsibility didn't make those kinds of mistakes."[10]

Bill Larson, of ABC Radio and a veteran of local Florida reporting, described that period in his career as a lifestyle. "Back then you could walk into one of the restaurants or bars or whatever, and the chances were you would run into one or two of the astronauts. You could sit and talk, have a beer. And it was that kind of a feeling amongst the people here; the astronauts were like members of the family."[11] Gene Cernan agreed. "I had a lot of good friends in the press. We invited them into our house under the condition that, once they're in our house, we have a beer or a scotch and soda, you leave the business outside."[12]

Naturally, that also made it difficult for journalists close to the astronauts to report on the personal side of their lives—the *Life*/World Book contract aside. "We got away with things that other people wouldn't get away with, whether it was speeding or going places we couldn't otherwise afford," Cunningham recalled. "One thing after another came our way and we didn't take the high road necessarily. We took advantage of a lot of those things. You couldn't go anyplace without all the women in the place looking at you like, all of a sudden, you were Superman. It was an unreal kind of existence because, during those days, we were the celebrities' celebrity."[13]

The reporters who covered the astronauts weren't blind to what was going on. Like their counterparts in Washington, D.C., they chose not to report it. There was an understanding among the establishment press of that time to respect the boundaries between personal lives and professional lives. "Of course I knew about some very shaky marriages, some womanizing, some drinking, and I never reported it," recalled Dora Jane Hamblin, one of the *Life* magazine reporters. "The guys

8. Walter Cunningham, op. cit., p. 164.
9. Doug Ward, interview with the authors, November & December 2012.
10. Ibid.
11. Bill Larson, interview with the authors, January 2012.
12. Gene Cernan, interview with the authors, February 24, 2012.
13. Walter Cunningham, JSC Oral History Interview, May 24, 1999.

wouldn't have let me, and neither would NASA. It was common knowledge that several marriages hung together only because the men were afraid NASA would disapprove of divorce and take them off flights."[14]

While *Life* wouldn't report these stories, it didn't mean that more sensational outlets wouldn't. The rumor mill was active and rampant. The astronauts, as celebrities, were exposed to all the temptations of modern day celebrity: groupies, hangers-on, wild parties, and handshake business propositions. Celebrity gossip magazines, such as *Confidential*, sent reporters to the Cape looking for stories, and some rumors couldn't escape the ears of the local Florida and Houston gossip columnists. But the mainstream press tended to avoid these stories. They were committed to covering the big story and remained as focused on the goal of reaching the Moon as NASA and the astronauts.

During the '60s and early '70s, astro-gossip remained on the fringe. "Nearly all [of the media coverage] had us squarely on the side of God, country, and family," wrote Buzz Aldrin in 1973. "To read those accounts was to believe we were the most simon-pure guys there had ever been. This simply was not so. We all went to church when we could, but we also celebrated some pretty wild nights."[15]

In his memoir covering the same era, Walt Cunningham added, "What wasn't realized at the time was the particular pedestal on which the media had placed us. In the years ahead they would contribute to protecting our reputations in a manner usually reserved only for national political figures."[16]

"In those early days, the PR people and the news people were very close and worked together," added Larson. "Everybody was so closely knit on one desire: to get to the Moon. It was an amazing feeling. I've never been anywhere before or since with that sense of camaraderie and a belief in that you were involved in the greatest exploration in the history of mankind. It was a cooperative effort on everybody's part."[17]

Mark Bloom, veteran reporter for the *New York Daily News* and Reuters, concurred. "It was a great adventure story. We wanted them to make it. I wanted them to make it. We were very happy that they made it. But they weren't perfect. We let a lot of the warts go by—I did. I let a lot of the warts go by."[18]

Conspicuously, one figure whose personal life generated little attention in the press was Neil Armstrong. Even when he divorced his wife of thirty-eight years, in 1994, the press barely noticed. Sometimes erroneously described as a recluse during the years after his retirement from NASA, Armstrong was not averse to making public statements and occasionally appearing at benefits supporting causes he championed. He even appeared in Chrysler television advertisements in 1979.

International celebrities, the Apollo 11 astronauts Neil Armstrong, Buzz Aldrin, and Michael Collins are swarmed by thousands of intense fans in Mexico City as part of NASA's global Presidential Goodwill Tour, which carried the astronauts and their wives to twenty-four countries over the course of forty-five days.

Armstrong and NASA management were keenly aware of what had befallen Charles Lindbergh in the years after his 1927 flight—the tragic kidnapping and death of his son, and the aviator's controversial statements on the eve of World War II—and it has been widely reported that Lindbergh's history influenced Armstrong's decision to maintain a low public profile. In fact, Armstrong and Lindbergh, two shy Midwesterners with much in common in addition to their sudden fame, struck up a friendship shortly after the return of Apollo 11 and became active pen pals.

Equally aware of Lindbergh's biography were the American journalists who had covered Armstrong's years at NASA. Their decision to respect the privacy of "The First Man," was continued by their successors during the last four decades of Armstrong's life. The small number of news items published about Armstrong's last forty years was both a reflection of Lindbergh's shadow on the history of celebrity and a lasting testament to the unique bond that formed between the press and the astronauts during the Apollo era. ⦿

14. Howard E. McCurdy, *Space and the American Imagination*, Washington, DC: Smithsonian Institution Press, 1997, p. 102.
15. Edwin Aldrin, *Return to Earth*. New York: Random House, 1973, p. 302.
16 Walter Cunningham, op. cit., p. 102.
17. Bill Larson, interview with the authors, January 2012.
18. Mark Bloom, interview with the authors, January 13, 2012.

It's understanding how passengers think that's helping us get more of them.

When you call Eastern Airlines Reservations, we know you want pleasant helpful service, and we know you're in a hurry.

So our goal is to answer your call quickly—and in most cases by a real person, not a recording.

Your call to Eastern Reservations is our first chance to take care of you, and that's exactly what we're going to do.

There are over 3,000 specialists in 12 Eastern Reservations Centers, handling more than 150,000 calls a day, one at a time.

Reservations is a big job. It's not just to sell seats. It's to answer questions about airfares, hotels, car rentals. It's to get you on the right flight even if we have to put you on another airline.

Simply said, Eastern Reservations people are there to help you.

It's this kind of attitude that helped win us 2 million more passengers last year than the year before.

Look, there are a lot of airlines out there you could fly besides Eastern. So if we want you to fly Eastern all the time, and we do, we have to earn our wings every day.

Every day. On the phone. At the airport. On the plane.

"The Wings of Man" is a registered service mark of Eastern Air Lines, Inc.

Introducing something else the space program launched.

When we say Rëvo® sunglasses are the best on the planet, we've got a good reason.

They're the only ones created by leaving it.

While working on a coating system to protect delicate lenses in space, a team of aerospace engineers discovered how to protect your eyes here at home.

The result is a revolutionary iridescent coating system more than 25 layers deep. A coating system that actually stops all harmful ultra-violet and infrared radiation, yet at the same time lets in much more usable light.

So you see with truer colors, better contrast and remarkable clarity.

Try a pair. Whether you wear sunglasses for driving, for sports or simply for spectating, there's only one way to describe their performance. Out of this world.

For more information, call (800) THE-REVO. Or in California, call (800) FOR-REVO.

Rēvo®

"Businessman, doctor or astronaut, we all need expert guidance in planning our financial security."

Jim Lovell, Astronaut.

With all he had to learn to fly to the moon twice, Jim Lovell is still quick to admit the need for expert guidance in other things.

And that includes planning for his financial security—today both as a family man and as a businessman.

Fortunately today, there is someone especially well qualified to give you this expert guidance.

Your Mutual Benefit representative.

He can help you no matter what form your desire for financial security takes. As a family man. As an incorporated professional. As a businessman requiring group life and health, pension or employee benefit plans.

Over the years, Mutual Benefit has earned a reputation for the quality of its policies, prompt payments and remarkable lack of red tape. The kind of service that has made it a favorite of accountants, lawyers, bankers—the financial professionals all over town.

So for expert guidance, talk to your Mutual Benefit representative. He can help.

Take it from Jim Lovell.

MUTUAL BENEFIT
You take care of today.
We'll take care of tomorrow.

"After our first Moon landing, I resolved I would never publicly endorse any product or service unless I sincerely believed in it."

Apollo 11 Astronaut, Buzz Aldrin, discusses the Computerized Self-Analysis system in every 1972 Volkswagen.

"Everything after a Moonwalk is a letdown.

You can't top it.

So instead, you adjust.

You adjust to a world that's a far cry from perfection."

Or at least the kind of perfection that took Buzz Aldrin to the Moon and then brought him back within a quarter mile of an aircraft carrier carrying the President of the United States.

Then it's over and there he is. With a Doctor of Science degree from MIT and a walk on the Moon under his belt, in a world, that at times, seems absolutely chaotic.

Then, something comes along that makes a great deal of sense to him.

And he becomes enthused:

"In analogy with the manned space flight programs, past, present and future—Mercury, Gemini, Apollo, Sky Lab and Space Shuttle—which pioneered computer check-out systems, the present Volkswagen system compares with the early Mercury flight of Shepard, Grissom and Glenn."

Volkswagen's new Computerized Self-Analysis system:

To Buzz Aldrin, it's a **"great step forward in the area of troubleshooting of automobiles."**

To you, if you own a 1972 Volkswagen, it means getting the most advanced automobile check-up there is.

"The car, truly, is wired along the same principle of a space craft."

(Nothing like asking the man who manned one.)

How the system will work at VW Dealerships:

In its simplest terms, it's a network of check points and sensors throughout the car; each sensor reporting the condition of various key points in the car to one central socket in the back of the car.

The socket, when plugged into a computer, which will be at VW dealers starting later this year, will actually report the condition of your car directly to you.

Things like your electrical system, cylinder compression and front wheel alignment will all be checked without human error.

Altogether, 60 vital service points will be checked. The results will then be printed out on a sheet of paper, in plain English, for you to read.

"Keeping that print-out sheet after each check-up is like checking-in with Mission Control when you're 200,000 miles out in space. It gives you a nice, secure feeling."

It also gets you home quicker.

"The technical advances of the system amaze me, of course, but the speed of operation amazes me even more."

To check cylinder compression the conventional way today, a mechanic has to do the following: remove each spark plug, insert a gauge, crank the engine, take a reading, make a decision and then put each spark plug back.

At top speed, the procedure takes 10 minutes and 40 seconds for four cylinders.

With the new VW system, it takes 60 seconds.

To check out your electrical system, you need a great deal of knowledge, equipment and 9 minutes and 40 seconds.

With the new VW system, your electrical system is checked out in less than 2 minutes.

At top speed, it takes the best mechanic using the best equipment 4 minutes to check out your front wheel alignment.

With the new VW system, elapsed time is 60 seconds flat.

So there you have it.

Instead of a mechanic telling you what's wrong with your car, now your car and your dealer's computer can tell you.

It took Volkswagen 7 years and enough money to develop and market an entirely new automobile to come out with the Computerized Self-Analysis system.

Why such a concentrated effort on service instead of, let's say, a new, longer, lower, wider Beetle?

Basically for the same reason Colonel Buzz Aldrin decided to publicly support it.

The reason is known as progress.

Postscript: Earlier this year, Volkswagen felt it had something advanced enough to show, not only to a man who walked on the Moon, but also a man who was intimately familiar with the subject of computerized check-out and testing.

*After a series of meetings and after seeing the system in operation and learning of plans to bring in the computers starting later this year, Colonel Aldrin's response, very simply, was: **"I'm impressed."***

WALLY SCHIRRA: One of the seven original American astronauts, retired Navy captain and test pilot—presently a business executive and industrial consultant—Life Member of the National Rifle Association.

"I've been in space three times and it always made me think about how good the Earth is. I guess that's why I never spend a day in the field when I don't think about the beauty of our country. And sometimes I get so absorbed with the environment I forget all about hunting. It's a great land and I'll take walking its ground over space travel any time."

"I'm a hunter and find it's good for the body and the mind. I've been a Life Member of the NRA for a number of years because Wally Schirra and his buddies want to be free to hunt next year and the years after that. No group works harder than the NRA to protect the sport and make hunting safer for everyone."

I'm the NRA.

The NRA introduced America's first hunter safety training course in 1949. Today, every state and Canadian Province provides hunter education courses modeled after the NRA program. If you would like to join the NRA and want more information about our programs and benefits, write Harlon Carter, Executive Vice President, P.O. Box 37484, Dept. WS-13, Washington, D.C. 20066.

Paid for by the members of the National Rifle Association of America.

Celebrate their odyssey into space at louisvuittonjourneys.com
Some journeys change mankind forever. Sally Ride, first American woman in space.
Buzz Aldrin, Apollo 11, first steps on the moon in 1969. Jim Lovell, Apollo 13, commander.
Louis Vuitton is proud to support The Climate Project.
LOUIS VUITTON

LEFT: *Many of the Apollo astronauts found career opportunities working or serving as spokespeople for major U.S. corporations following their service. They were media stars of the era and many companies and associations wanted to identify with their heroic status and tech-savvy, all-American image.*

ABOVE: *In celebration of the 40th anniversary of the first moon landing, this 2009 Louis Vuitton ad, created by Ogilvy & Mather, featured three of the best-known astronauts—Sally Ride, Buzz Aldrin, and Jim Lovell—in a photograph by Annie Leibovitz. Like the old and weathered truck, the astronauts had lost none of their ability to personify the spirit of adventure. They were still an item and would always be.*

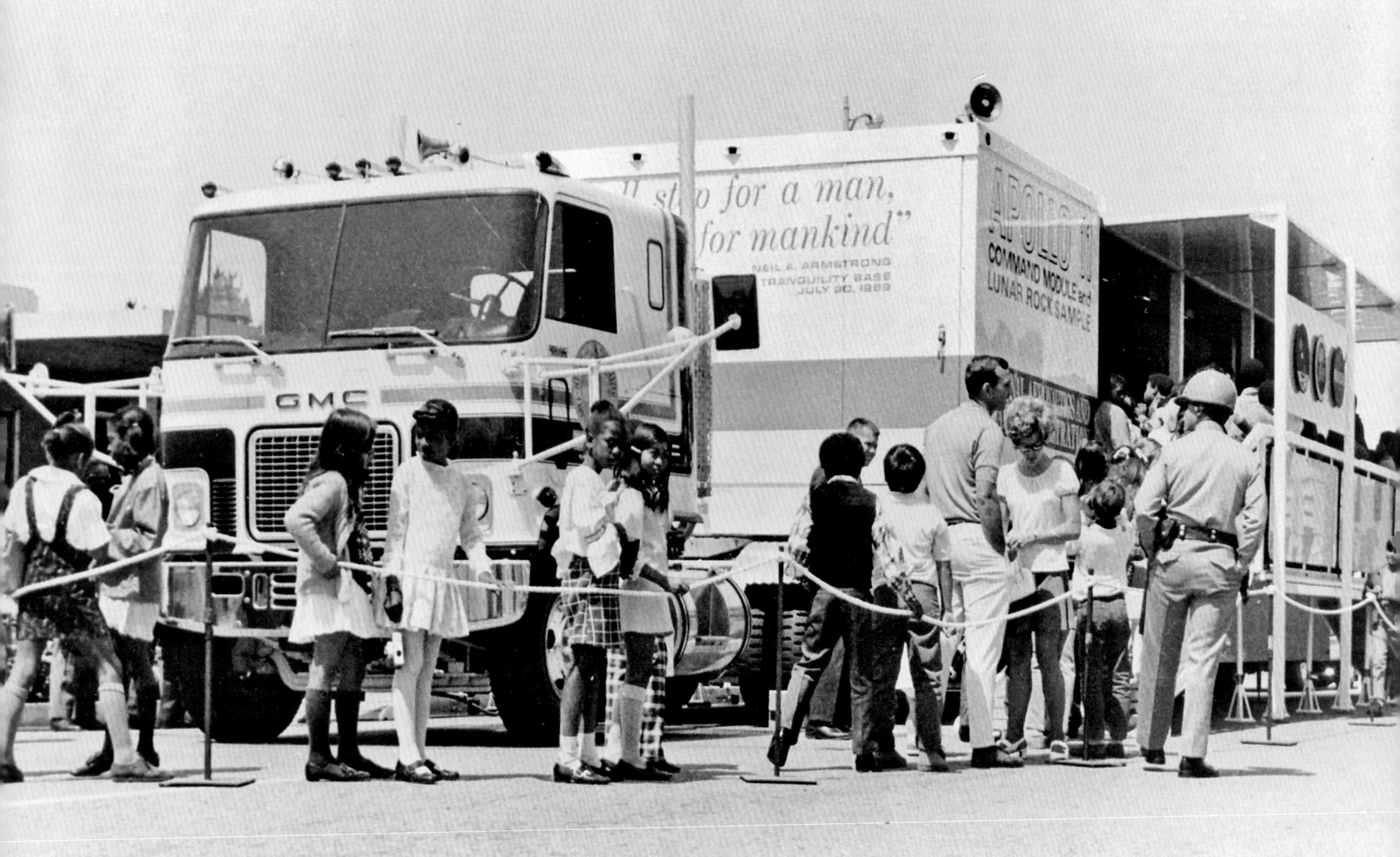
for mankind”
NEIL A. ARMSTRONG
TRANQUILITY BASE
JULY 20, 1969
APOLLO 11
COMMAND MODULE and
LUNAR ROCK SAMPLE
GMC

The Apollo Roadshow: Moonwalkers and Moon Rocks

"From Space to Grass-roots America"

—NASA's slogan for the Apollo 11 Fifty-State Tour

By the end of 1969, the American public's interest in the Apollo program was starting to wane. Despite the hard work of NASA's Public Affairs officials and many others who contributed to the program, the results of public relations efforts after the Apollo 11 mission were decidedly mixed, lost in the inevitable down-draft of enthusiasm after Kennedy's audacious goal was decisively achieved. Now the eyes of the country turned earthward, as the social and political problems of the day regained the lead headlines.[1] A number of news articles and editorials noted this rapid shift in national sentiment, often lamenting the fickle, yet hardly surprising, state of the collective attention span. An April 1970 editorial in the *Playground Daily News*, a Fort Walton Beach, Florida, daily newspaper, pointedly articulates the troubled Apollo *zeitgeist*:

> Nothing succeeds like success—in dulling the public's interest. Some 3,000 reporters and photographers attended the launching of Apollo 11 on its history making flight last July 16. Hundreds of millions of people around the world followed the saga on their television screens. In November, 1,700 newsmen witnessed the lift-off of Apollo 12. A week before the scheduled April 10 launch of Apollo 13, for the third manned landing on the Moon and the most ambitious lunar exploration assignment thus far, only 900 requests for press credentials have been received by NASA. Journalists aren't the only jaded ones. According to *Aviation Week & Space Technology* magazine, the world tour by the Apollo 12 crew is being called a public relations flop by some NASA officials, who are arguing against a similar trip by the Apollo 13 astronauts after they return to Earth. Crowds at parades and receptions for the three Apollo 12 crewmen have been noticeably smaller and less enthusiastic than those during the tour of the Apollo 11 crew. Regrettable as it may be, this decline of popular interest in the flights of Apollo 12 and 13 was to be expected. How many times can people flip over the sight of an astronaut or a chunk of rock, which, one has to keep telling oneself, actually once rested on the Moon? Nothing can, and possibly ever will, duplicate the excitement that surrounded an event unique in all human history—certainly not virtual reruns of that event."[2]

In the face of this wave of declining interest, the NASA Public Affairs Office tried a number of different strategies to reignite enthusiasm, while striving to position the country's recent achievement and Apollo's scientific discoveries forefront in the national consciousness. This even included taking *Columbia* on the road like a traveling rock band promoting an album.

"Every opportunity we could come up with to tell the story, to tell the nation where that less-than-one-penny-on-the-dollar went, we wanted to tell that story," recalled Chuck Biggs, NASA Public Affairs officer from the Manned Spacecraft Center in Houston, and the head of exhibits and tours for NASA during the Apollo program.[3]

Throughout 1970 and 1971, NASA conducted an unprecedented Apollo 11 Fifty-State Tour featuring the command module, an Apollo 11 Moon rock, and other artifacts that had flown on the mission. The tour was a national, grass-roots roadshow intended to bring the historic craft and the story of its technical achievement to the citizens of each state, both as a thank-you, as well as a direct pitch to the public for the benefit of future manned space exploration. While unprecedented in the history of the Apollo program, *Columbia*'s Fifty-State Tour was inspired in part by 1962's international traveling exhibit displaying John Glenn's Mercury *Friendship 7*, and a tour of American state capitals of Gordon Cooper's Mercury *Faith 7*, in 1963 and 1964.

"Headquarters decided we should take the Apollo spacecraft around to every state capital," Biggs explained. "And they took the lead in designing the tour. But as they didn't have the capability to design and build such a vehicle that would move

***OPPOSITE:** People in line at the Nevada state capital at Carson City to view the Apollo 11 spacecraft, Moon rock, and other display items as part of the NASA Fifty-State Tour in 1970.*

1. In a 2003 article, "Public opinion polls and perceptions of US human spaceflights," published in *Space Policy*, historian Roger D. Launius argues that, contrary to perceived belief, there is little evidence of American public enthusiasm for Project Apollo, except during the summer of 1969. Lanius writes, "consistently throughout the [1960s] 45–60% of Americans believed that the government was spending too much on space, indicative of a lack of commitment to the spaceflight agenda."
2. Don Oakley, "Moon Shots Lose Whiz-Bang Aura." *Playground Daily News*. April 13, 1970, p. 4.
3. Chuck Biggs interview with the authors, October 17, 2011.

Official NASA Apollo 11 Fifty-State Tour press photo of the GMC van, sponsored by the Heavy Specialized Carriers Conference of the American Trucking Association, and its official escort cars on the road in May of 1971.

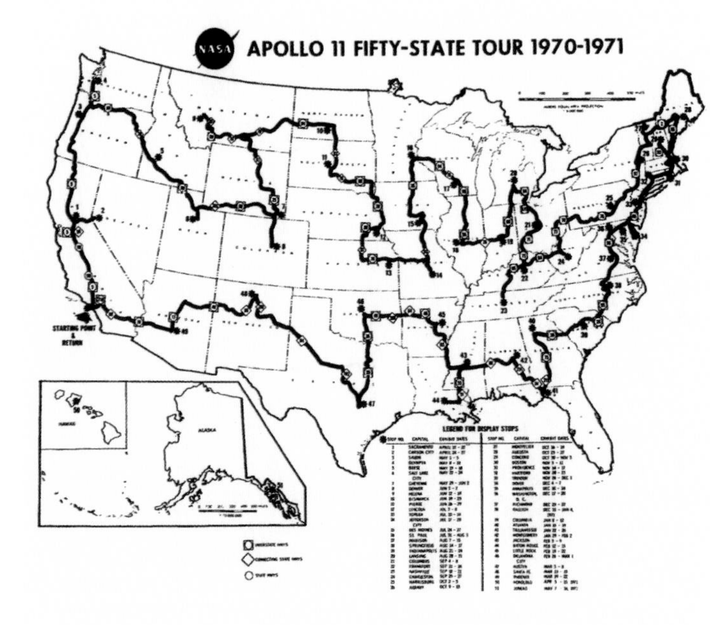

Map of the tour route from the NASA press kit.

an Apollo capsule around, they asked my office to do that. I relied on Colin Kennedy, who was an excellent designer in the office, and a couple of our contract people."[4]

"In October of 1969, Chuck was charged by NASA headquarters to design the van and the exhibit," remembered Elwood Johnson, an experienced NASA tour exhibitor and educator, whom Biggs hired as the Fifty-State Tour director. "Chuck and his team were very well respected. I remember visiting him and Colin Kennedy at the Ramada Inn in Houston. They designed the van on an envelope or a napkin! Those guys were brilliant."[5]

"The van itself was brand new," Johnson said. "It was a GMC truck. I thought, given our cargo, that we'd only go about thirty-five or forty miles-per-hour, but we often got up to seventy and eighty miles-per-hour down the road on the interstates and highways with our police escort. If cars got in the way, the state police would use their sirens. Can you imagine seeing this thing coming up behind you in the rear view mirror? It would scare the hell out of some drivers."

Even at seventy or eighty miles-per-hour, its speed paled in comparison to the more than 24,000 mph speed the spacecraft reached during some of its million-mile journey during the mission—a fact often emphasized in the NASA press materials throughout the tour.

"Although NASA's Apollo 11 Command Module moved at a much more sedate pace in its visits to fifty state capitals than its historic journey through space, its earthbound trip was equally successful," wrote Willis H. Shapley, NASA Associate Deputy Administrator at the time of the tour. "The Space Agency's gratification in affording 3.25 million Americans a close-up view of Apollo 11 and its accompanying Moon rock display was more than matched by the pride and enthusiasm of the citizens welcoming it in each state."[6]

By this time, NASA was already enduring considerable program budget cuts after the goal of reaching the Moon had been achieved. Congress and the Agency were struggling with the need to balance space expenditures with other pressing national priorities, something that was hardly lost on Shapley, who also commented: "We take special satisfaction in noting that the tour was a commendably efficient one, from the standpoint of the numbers of personnel and dollars involved. We were particularly pleased with the public-spirited response of

4. Chuck Biggs, JSC Oral History Interview. August 1, 2002.
5. Elwood Johnson, interview with the authors, June 11, 2012.
6. Donald Zylstra, NASA Report: The National Aeronautics and Space Administration's Apollo 11 Fifty-State Tour, 1970–1971. Submitted to NASA June 1971.

the [independent trucking companies and] associations, who made the extended trip possible without charge as a public service."

"You know, NASA was criticized over the years for not doing more, but we did a lot, and as Public Affairs," explained Biggs. "Admittedly, we (often) responded to the demands, but once we realized the demand was there, I think the agency did a darn good job in providing information and experiences to the public."

"Washington did all the scheduling with the various states and governors, and they set up all the transportation," Johnson recalled. "We had three drivers—one for the trailer, and two for the escort vehicles. And they would change with every state. I didn't have to worry about the publicity, or the documentation—the photographs or even counting the visitors. That was all handled by each individual state governor's office."

The tour, and its arrival in the individual fifty capitals, received considerable news coverage in the local press, but seldom garnered national press attention—partly the result of dwindling public interest, but primarily due to the fact that this was a regional news story for which most publicity was handled locally.

"That was probably a mistake on somebody's part," said Johnson.

In addition to the command module, the tour van also included one of the first publicly exhibited Moon rocks recovered at Tranquility Base a few months earlier. It was stored inside the van for the entire tour. "We had a safe that was welded to the floor in front of the van," recalled Johnson. "Every morning I would make sure that everything was clean and opened up for the day's crowd, and I would take the Moon rock out of the safe and brought it into the van proper. It would stay there, on display, all day, until when I closed up between 8:30 and 9:00 P.M. at night. Then I would lock the Moon rock back up in the safe."

"Of course, we had state police for security twenty-four hours a day," said Johnson. "We had a state police escort when we were in transit, and we had state police at the exhibit site as well. This was in the 1970s—and it was really during a time of student riots. In the country, you know, they were disrupting things, and the police and others were very worried about

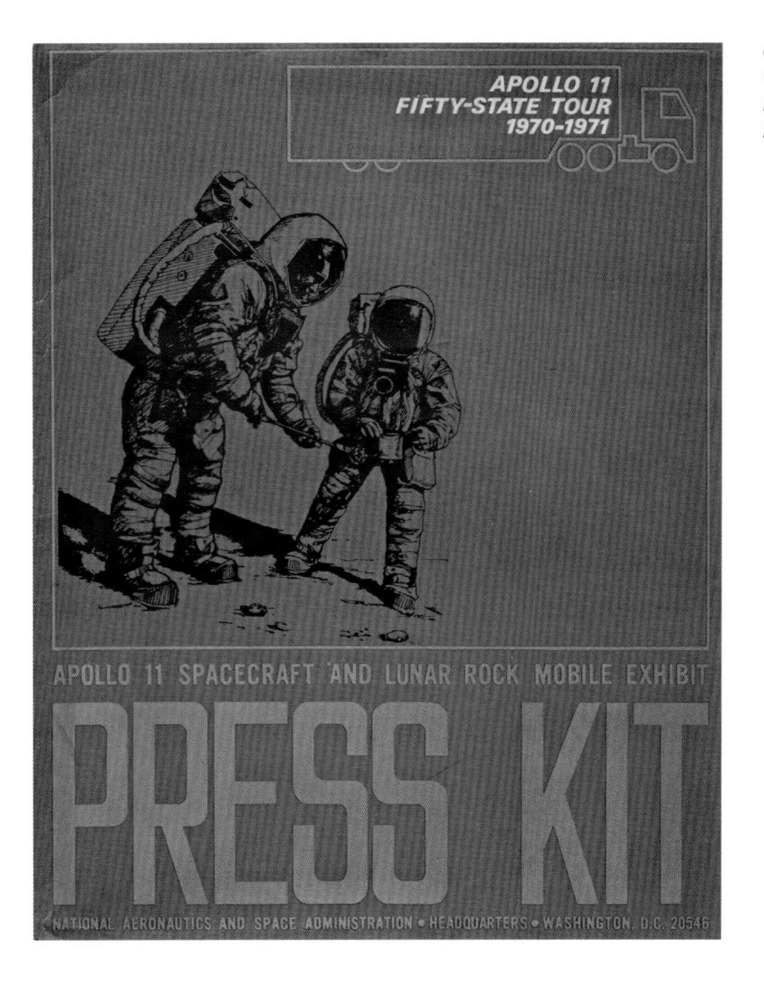

Cover image of the official NASA Apollo 11 Fifty-State Tour press kit folder, which included a design rendering of the layout of the van.

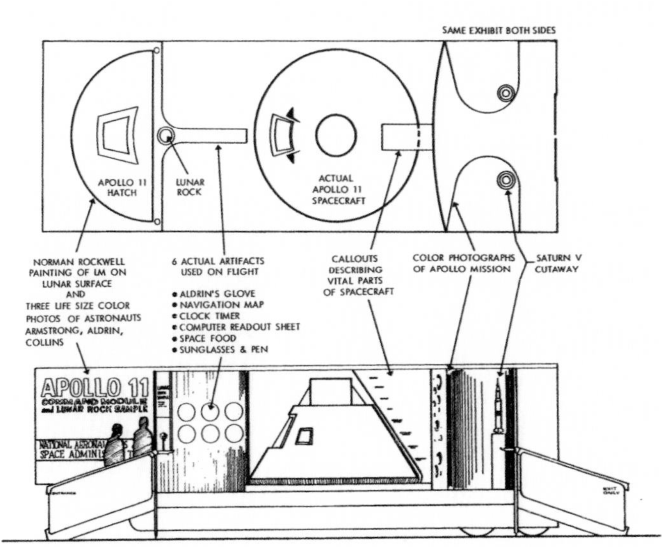

Apollo 11 one-year anniversary celebration at the tour exhibit site in Jefferson City, Missouri, on July 20, 1970. It was the first time the crew had seen the spacecraft since returning from the Moon. In the photo are Missouri Governor Warren E. Hearnes and Neil Armstrong.

that. And you know what? We didn't have any issues—not one." Nevertheless, NASA was careful enough to insure the capsule and van for $10 million, just in case.[7]

Despite the major effort and cooperation of private industry and government during the tour, there was usually only one, lone NASA Public Affairs employee who accompanied the tour to each state, meeting with the governors and dignitaries at each stop. It was a grueling schedule, and Johnson made it through thirty-three of the fifty states, before the breakneck schedule took its toll.

"I went to thirty-three states, and then I ran out of steam," said Johnson. "But I think the public got a lot out of the tour, and a lot of that was a sense of national commitment and pride. It was absolutely amazing what this country could do."

The tour commenced in Sacramento in April 1970, and included visits from NASA dignitaries and then-Governor Ronald Reagan. Johnson spent his mornings opening up the van, and then changed into a suit to greet the governor and other dignitaries that showed up for the event. Notably, Hawaii drew the greatest number of exhibition attendees—135,000—attributable in part to the state's visiting tourist population and its proximity to Asia. The second highest attendance was reported in North Carolina, where 100,000, many of whom were school children, stood in line to see *Columbia* and contemplate a tiny gray chunk of the Moon. Maryland had the fewest, 32,500, followed closely by Washington, D.C. with 35,000 (despite a visit from Neil Armstrong). Consistently larger draws were experienced in the states where major NASA contractors worked on the Apollo program, or in which NASA had a major research lab or facilities presence: California (91,000), Massachusetts (90,000), Texas (90,000), Alabama (89,000), Ohio (80,000), and Florida (70,000). The most amazing public reception, given a location's population density and weather would have to be Alaska, where 96,000 visitors responded to extensive TV, radio, and newspaper publicity, and the fact that it was the final stop on the tour, which had been extended for eleven days.[8]

After waiting in line, attendees entered the truck where they were able to closely examine the command module, its interior protected against possible damage or defacement by a thick, clear plastic sheath. "It's smaller than I thought," remarked one amazed visitor in New Mexico, as enthusiastic visitors were allowed to gaze on the capsule's weathered exterior, and trace their fingers over the pock-marked impacts of micrometeorites and the intense heat burns to the outer skin of the craft.[9]

"The truck visited a state for five days, and then went on to the next state," recalled Johnson, explaining the logistics of his one-van show. "We came into a state, say, on a Thursday, and that's when I set it up. I would take down the sides for the flooring and the ceiling. It was built into the van, like a camper. And it took me about five hours to set that thing up. So Thursday, we set it up. Then we opened up on a Friday, and we were there Saturday and Sunday, and maybe, sometimes, Monday. Then I had to close it all down—which took a lot less than five hours to do—and then we'd travel on to the next state, getting to the capital on Thursday, and I would set it up all over again."

It was often reported that visitors regularly queued up in lines that were several blocks long, with most people waiting hours for their opportunity to enter the exhibit. Occasionally,

7. Elwood Johnson, interview with the authors, June 11, 2012.
8. Donald Zylstra. NASA Report, op. cit, p 29–30.
9. "Crowds Flock to Apollo 11." *The Sunday New Mexican*, March 14, 1971, p. 1.

an astronaut would visit the exhibit—Buzz Aldrin opened the exhibit when it arrived in Austin, Texas—but most stayed only a few minutes. The majority of NASA's featured speakers, a total of twenty-five during the entire program, were chosen from its research centers and laboratories, and usually spoke at host luncheons or special evening events.

The only big-draw astronaut visit was on the first anniversary of the first Moon landing, when the crew of Apollo 11 made a brief visit to the exhibit in Jefferson City, Missouri. Also attending that event were NASA Administrator Dr. Thomas Paine, and Julian Scheer, the Assistant Administrator for Public Affairs. The anniversary event, highlighted by the attendance of the Apollo 11 crew, drew an estimated crowd of 7,000.[10]

"That was the big one, with all the big guns there," explained Johnson. "But for the most part, the astronauts would come out and they would be at the exhibit probably for about five minutes, nothing more, and then would leave." In Indiana, Gus Grissom's family came out, and Neil Armstrong's brothers attended the opening in Indianapolis. "Most people at each stop were very nice and pleasant, and we didn't really have any trouble. After a while, it just became very routine," Johnson recalls.

Even though the display van that carried the exhibit was designed by Biggs and his team, the exhibit itself was assembled under a NASA contract by Industrial Displays, Inc.[11] Pulled by a GMC truck cab, it measured fifty-eight feet long, fourteen feet wide, and thirteen feet tall. Transportation costs throughout the Fifty-State Tour were covered by the Heavy Specialized Carriers Convention of the American Truckers Association, as well as by Matson Navigation Company, and the Sea-Land Service company—a fact NASA chose to emphasize with each press release and press photo. More than 125 drivers and the services of over seventy-eight member firms of the Heavy Specialized Carriers Convention contributed to the success of the program, but it was Biggs, Johnson, and the MSC Public Affairs team that got the wheels moving across the U.S.

Reflecting more than forty years after his marathon tour across the country with *Columbia*, and the millions of people who walked through the exhibit, one special visit remained vivid in Johnson's mind.

"We were at the Ohio state fairgrounds," Johnson recalled, where more than 80,000 people visited the exhibition during three days in September of 1970. "And on the first day, someone came up to me and said, 'You know what? Neil Armstrong's parents are at the back of the line!' And I couldn't believe it. They were going through the line like everyone else, just waiting patiently. So I walked up there and introduced myself. I had never met them before, but they were just the nicest people in the world. They felt like your own mom and pop. I mean, they were absolutely wonderful."

"People around us, of course, didn't know who they were. So I took them up, brought them to the capsule, and through the van. They were very excited, especially to touch the capsule, but they didn't ask any questions. They found it difficult to say much of anything, because there was a lot of emotion." Johnson paused during his recollection, then continued: "They did not stay long, because they were cognizant that they weren't in front of the line, that I had brought them out in front of the other people. They didn't want the people in back to wait because of them. They were just amazing, salt-of-the-earth-type people."

Long lines queuing up for a view of the Apollo 11 command module at Carson City, Nevada, April 1970. A total of 43,750 people visited the exhibit van during its three-day stay there. According to the post-tour NASA report, attendance was considered very good for Nevada, given the state's population. School children were bused in from around the state to visit the display, and "excellent press coverage prompted thousands of adults and their families to journey long distances to see it."

10. "Crew Visits Columbia on Moon Landing Date." United Press International, July 21, 1970.
11. David Swaim, "Moon Rocks Given Dual Area Viewing" *Pasadena Star-News*, April 15, 1970, p. 1.

(HO2)SPACE CENTER, Houston, Tex., Oct. 29--MOON GEM ON DISPLAY--Crowd at left moved in close to the white sphere to get a look at the one ounce piece of moon rock that went on display for the general public. The one ounce rock sparkled with many imbeded crystals and was backed up by a small replica of the moon. (APWirephoto)(see APWirestory)efk41645stf-efk1969

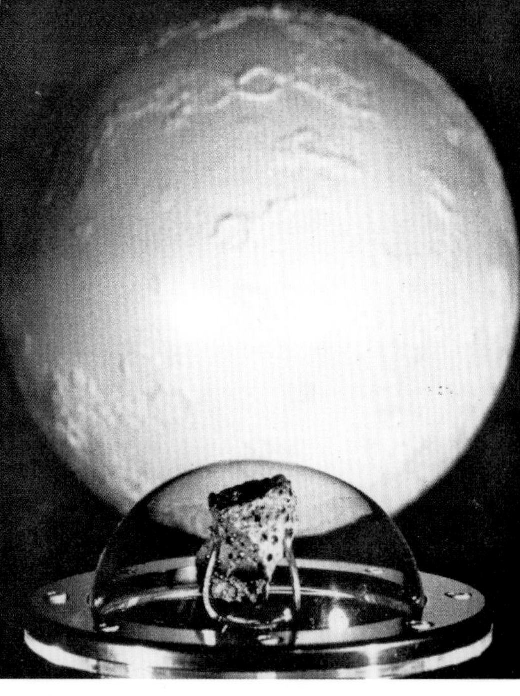

The tour, which covered more than 26,000 miles—14,000 by land, and 12,000 by sea—concluded its trek first in Honolulu and then, finally, in Anchorage, Alaska, the only non-capital stop. Through more than a year on the road, NASA calculated the exhibit drew roughly 3.25 million visitors, a number at least one million greater than NASA's original estimate in its 1970 press kit announcing the tour.[12] By that measure alone, the tour was a success. However, the number was a small fraction of the tens of millions of people who tuned in to see the launch live on TV, and just a little more than a fifth of the people who visited the U.S. pavilion lunar rock exhibit at Expo '70 in Japan. Despite the best efforts of NASA PAO, it was clear that, by the early 1970s, America's appetite for the lunar space program had changed.

Rocks on Tour

It was opening day of Expo '70 in Osaka, Japan, and a local security guard was in a panic. "Please help us," he pleaded with an American official at the United States pavilion as more than 8,000 people pushed their way into the exhibit hall. "We don't have enough men to stop them!"[13]

Luring the surging crowds to the 100,000 square-foot United States pavilion on March 15, 1970, was neither the enticement of immediate riches nor a celebrity of world renown. Rather it was the rare opportunity to gaze upon actual vehicles from the American space program and to get as close as possible to a truly unearthly fragment of the universe. The spectacular achievement of Apollo 11 less than a year earlier was celebrated with a vast lunar landing panorama around which were displayed the Apollo 8 command module, an actual Moon rock returned by Apollo 12, and a number of other significant artifacts tracing the history of the United States's ventures into space.

A main attraction at the first World's Fair ever held in Asia, the United States pavilion and its singular Moon rock exhibition welcomed an average of 80,000 visitors per day, and was seen by at least 14 million before Expo '70 closed six months later, in September. (On peak days, waiting in line to enter the exhibition might take as long as two hours.)[14]

Between 1969 and 1972, the six Apollo lunar landing missions returned to Earth exactly 2,196 individual Moon rock, soil, and core samples, representing 842 pounds of ancient and

FAR LEFT: Malcolm Toon, the U.S. Ambassador to Czechoslovakia, speaking at the opening ceremony of a scientific conference near Prague in 1970 with an Apollo Moon rock on display.

RIGHT, ABOVE AND BELOW: Photos of the U.S. and Soviet exhibit spaces at EXPO '70 in Osaka, Japan.

precious lunar material.[15] With all of that material on hand, NASA headquarters decided that some of it would be earmarked for public relations, rather than confined for research, scientific, and educational purposes.

One of Chuck Biggs's responsibilities was working with the United States Information Agency, which oversaw the planning of the United States pavilion in Osaka. "It was really in response to public demands, public needs, for the display of the lunar samples," Biggs recalled in a NASA oral history. "People wanted to show Moon rocks. So we developed a Lunar Sample Display Program. My boss at the time [Brian Duff] asked me if I'd work with the curatorial staff and do that. So we did . . . and then we maintained control of the lunar samples as they traveled throughout the world."[16]

Elsewhere in Osaka, the Soviet Union also had a pavilion of major prominence, in which the USSR's space program was accorded much attention. Near Tchaikovsky's piano was an impressive array of life-size space hardware, including Yuri Gagarin's original *Vostok 1* and twin-docked Soyuz spacecraft suspended from the ceiling. Large lines queued here as well, and visitors included the Apollo 12 astronauts, who came to Osaka a few days after Expo '70 opened. Pete Conrad, Alan Bean, and Dick Gordon toured the red and white building—the tallest at the Expo—and explored the Russian cultural and technological displays. But the clear distinction between the Soviet pavilion and the United States pavilion was the presence of the Moon rock, a symbol of America's victory in the race to land a man on the Moon.

Indeed, the Moon rocks proved popular the world over. NASA reported that by the end of 1970, more than 41 million people had viewed an Apollo 11 or 12 Moon rock. Three quarters of this number were people from 110 countries who saw lunar samples sent on tour by the United States.[17]

In addition to public displays and exhibits at museums, state fairs, NASA facilities, and global events such as the World's Fair, NASA also distributed Moon rocks and dust samples to scientists and research organizations around the world. After screening thousands of lunar sample requests from university, industrial, and government laboratories, the first distribution of lunar matter took place in September 1969. This alotment, eighteen pounds of rock and dust (by weight,

12. Official NASA Fifty-State Tour Press Kit, 1970.
13. "Moon Rock is Socko," *JSC News Roundup*, March 27, 1970, p. 1.
14. Ibid.
15. "NASA'S Management of Moon Rocks and Other Astromaterials Loaned For Research, Education, And Public Display." NASA Office of Inspector General. 8 December 2011, p. 1
16. Chuck Biggs, JSC Oral History Interview, August 2002.
17. "Over 40 Million See Moon Rocks." *JSC News Roundup*, February 26, 1971, p. 2.

roughly a third of the geologic samples returned by Apollo 11), was divided among more than 140 investigators in the United States and eight other countries. Five months later, nearly 200 scientists around the world got their chance to work on samples returned from the Apollo 12 mission.[18]

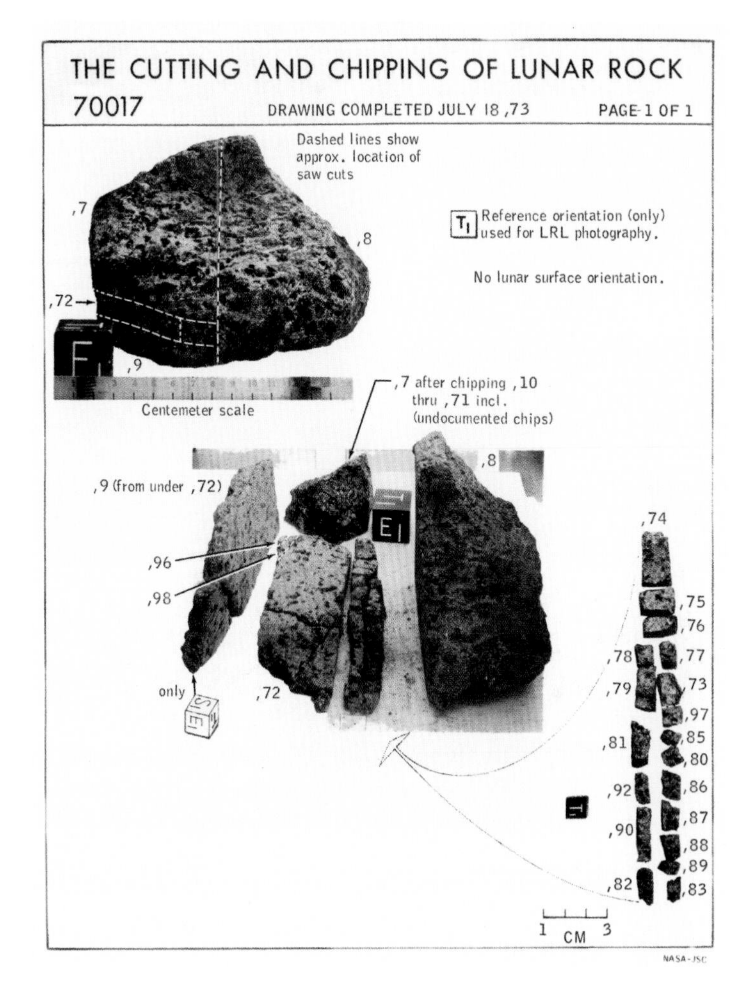

Diagram for the cutting of the Goodwill Moon Rock, Apollo sample #70017.

A Few Rocks and Samples Go AWOL

With so many samples in circulation, it was only a matter of time before some lunar material was reported missing, causing a public relations snafu a few weeks after the Apollo 12 distribution in February 1970. Without authorization from NASA, scientists at the University of California, Los Angeles, placed a vial of Apollo 11 Moon dust—2.3 grams from lunar sample rock number 50—on display in a department store during a $100-a-plate fund raising dinner. At some point in the evening, the vial was stolen. As news of the missing vial became public, the press questioned NASA's security protocols and its existing policy for the disposition of the Moon rocks. When asked by the FBI to place a value on the stolen vial, a NASA spokesperson from the Manned Spacecraft Center declined to name a specific price, but did point out that "it cost $23 billion to get it."[19]

The extensive press coverage had one positive effect: the sample vial was recovered within forty-eight hours, after an anonymous caller phoned the LAPD with information that the vial could be recovered at a designated mailbox.[20]

Other samples began to disappear as well—one from the Goddard Space Flight Center in July 1970, and two packages containing samples sent by NASA via the U.S. mail the following October. Dwindling press interest kept those stories from gaining wide distribution. Meanwhile, NASA made a concerted effort to restrict any additional unauthorized displays.[21] Maintaining tight control of Apollo's geologic legacy and lost samples continue to plague NASA into the 21st century, but it seldom merits a mention in the news media.

The Goodwill Moon Rock

In the years immediately following the lunar landings, the United States gave Moon rock samples as gifts to countries, heads of state, and to each of the fifty United States—often with great pomp and circumstance. Vice President Spiro Agnew generated extensive headline news when he bestowed Apollo 11 Moon rock samples to ten nations during a tour of Southeast Asia and the Pacific in December 1969.[22] An Apollo 11 Moon rock given to Mexico by United States Ambassador Robert McBride went on a national tour in 1971, during which it attracted more than 30,000 visitors on a single day in Chihuahua City.[23] And in early 1971, rival New Mexico politicians conducted a public battle over the possession of a Moon rock given the state. The outgoing Republican Governor, David Cargo, who had received the rock from President Nixon, refused to give up possession to his successor, Democrat Bruce King. "Those rocks were presented to me, not him. I can do whatever I want with them," Cargo was reported to have said by the press. Eventually, the rock was placed in a New Mexico museum.[24]

The success of the traveling Moon rocks, and the global demand for them, eventually led to the "Goodwill Moon Rock,"

18. NASA Press Release, Manned Spacecraft Center, Houston, TX, #MSC70-24, February 13, 1970.
19. Paul Recer. "Moon Dust Samples Are Sent by Mail." Associated Press, March 2, 1970.
20. NASA Press Release, Manned Spacecraft Center, Houston, TX, #MSC70-33, March 16, 1970.
21. NASA Press Releases, Manned Spacecraft Center, Houston, TX, #MSC 70-77, July 1, 1970, and October 20, 1970.
22. "Agnew to Carry Moon Rock Gifts," *The New York Times*, December 25, 1969, p. 1.
23. "Juarez Prepares Moon Rock Show," *El Paso Herald-Post*, May 6, 1971, Sec. D, p. 2.
24. "Moon Rock Ownership Disputed," *The Daily Mail* (UK), February 8, 1971, p. 11.

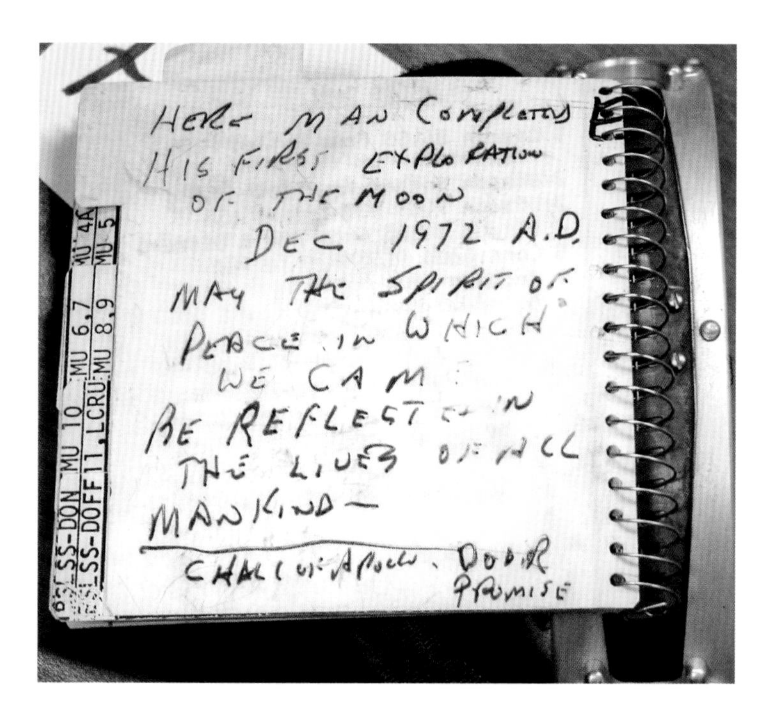

Astronaut Gene Cernan's notes for his Goodwill Moon Rock speech, delivered while standing on the surface of the Moon.

a program devised not by NASA Public Affairs or headquarters in Washington, but by the astronauts. In 1972, during the final Moonwalk of Apollo 17, astronauts Gene Cernan and Jack Schmitt made a deeply moving dedication of a rock sample to the people of the world.

"We understand there are young people (from countries all over the world) in Houston today who have been effectively touring our country," Cernan said on live television while standing on the surface of the Moon. "They had the opportunity to watch the launch of Apollo 17. Hopefully (they) had an opportunity to meet some of our young people in our country. We'd like to say, first of all, 'Welcome: we hope you enjoyed your stay.' Second of all, I think probably one of the most significant things we can think about when we think about Apollo is that it has opened for us—*for us* being the world—a challenge of the future. The door is now cracked, but the promise of the future lies in the young people, not just in America, but the young people all over the world learning to live and learning to work together. In order to remind all the people of the world in so many countries throughout the world that this is what we all are striving for in the future, Jack has picked up a very significant rock, typical of what we have here in the valley of Taurus-Littrow. The rock is composed of many fragments of many sizes, many shapes, probably from all parts of the Moon, perhaps billions of years old. A rock of all . . . sizes and shapes, fragments of all sizes and shapes, and even colors that have grown together to become a cohesive rock, outlasting the nature of space, sort of living together in a very coherent, very peaceful manner. When we return this rock and some of the others like it to Houston, we'd like to share a piece of this rock with many of the countries throughout the world. We hope that this will be a symbol of what our feelings are, what the feelings of the Apollo program are, and a symbol of mankind that we can live in peace and harmony in the future."

"A portion of a rock will be sent to a representative agency or museum in each of the countries represented by the young people in Houston today," added Schmitt. "We hope that they will—that rock and the students themselves will—carry with them our good wishes, not only for the new year coming up but also for themselves, their countries, and all mankind in the future. Put that rock in the bag, Geno!"

Cernan responded, putting what would become known as The Goodwill Rock, Apollo lunar sample #70017, in his large sample collection bag, saying: "I salute you, promise of the future."[25]

In early 1973, President Richard Nixon made good on Cernan and Schmitt's lunar request, decreeing the distribution of fragments from the rock be distributed to foreign heads of state, and to the governors of the fifty United States and its provinces. The Goodwill Rock presentations include a chip of lunar sample #70017 encased in Lucite, and a flag of the country or state that had been flown to the Moon. They remain lasting, symbolic gestures of peace from the end of a grand space program born of the immense international tensions and fears of the Cold War. ⦿

25. Apollo 17 Lunar Module On-Board Voice Transcription. Publication #MSC-07630, January, 1973, p. 374.

"So, Where to Now, Columbus?": The End of the Apollo Era

"It was the press who were saying, 'We've been to the Moon, so what are we going to do now? We've done that, and that was pretty easy. So, where to now, Columbus?'"

—Gene Cernan, the last man on the Moon

FEW PEOPLE ALIVE on December 14, 1972, can tell you where they were on that day. On the Moon, Gene Cernan and Jack Schmitt ended their final exploration at Taurus-Littrow and departed from the lunar surface in the lunar module *Challenger.* During the final moments of the third and last EVA, Cernan reflected on the experience, prefacing his comments with an aside that he believed it wouldn't be too long before humans returned to the Moon. "America's challenge of today has forged man's destiny of tomorrow. . . . We leave as we came and, God willing, as we shall return, with peace and hope for all mankind."

Minutes earlier, as he returned to the lunar module, Schmitt had directly addressed NASA Administrator Dr. James Fletcher, who was at Mission Control in Houston. As he looked at the dramatic landscape surrounding him, Schmitt noted that "This valley of history has seen mankind complete its first evolutionary steps into the universe: leaving the planet Earth and going forward into the universe. I think no more significant contribution has Apollo made to history. It's not often that you can foretell history, but I think we can in this case. And I think everybody ought to feel very proud of that fact."

Back on Earth, attentions were focused elsewhere. The final voyage of Apollo wasn't leading many of the evening newscasts, despite the fact that the astronauts were literally walking on the lunar surface and transmitting television signals at that very moment. A general mood of complacency, if not fatigue, regarding the Apollo program could be heard throughout the country and this was reflected in the television news coverage of the time. The evening of December 14, after showing a taped replay of the lunar module's liftoff, David Brinkley on NBC offered a brief commentary on Apollo, NASA, and the future of man in space that captured the *zeitgeist* at the moment the Apollo came to an end.

After praising Project Apollo for a job well done, Brinkley questioned why NASA wasn't being shut down and the money spent on something else. He described the future Space Shuttle program as a project with a vague and uncertain cost and purpose, and suggested it was a typical Washington boondoggle designed to perpetuate NASA's payroll and create aerospace jobs. "The American people might wonder," Brinkley concluded, "if all these billions and all this science, engineering and work might not produce something more useful; if this country really needs another expensive piece of hardware in orbit, when here on the ground we can hardly get the mail delivered. The Space Shuttle will take enormous amounts of money, talent and energy. In a country with problems as serious as ours, there must be some better way to use it."

Minutes after Brinkley's commentary was broadcast, the Apollo 17 crew had their own confrontation with the fading dream of Apollo. While the crew was securing the *Challenger* and transferring cargo into the command module *America,* capcom Gordon Fullerton read a prepared statement from President Richard Nixon to the crew celebrating their achievement. In it Nixon had noted, "This may be the last time in this century that men will walk on the Moon, but space exploration will continue."

Aboard the two docked spacecraft in orbit around the Moon there was concealed anger at the President's suggestion that this might be the final lunar mission of the century. Schmitt recalled his reaction in a NASA oral history recorded in 1991. "I thought that was the stupidest thing a President ever could have said to anybody. You may believe it privately,

OPPOSITE: Called "the most influential environmental photograph ever taken," "Earthrise" was the name given by its photographer, astronaut Bill Anders, who captured this image on the Apollo 8 mission, December 24, 1968.

No visual representation better captured the idea of the space race than this iconic cover illustration by Robert Grossman that appeared on the December 6, 1968, cover of Time *magazine. Published just prior to the mission of Apollo 8, when the world press believed the United States and the Soviet Union were in a virtual dead heat, less than a month later it was universally conceded that the United States had pulled far into the lead.*

but why say that to all the young people in the world. . . . It was just a totally unnecessary thing for him to say. Whoever wrote that speech really blew it with that remark. And I was really upset. Tired, but really mad. It was just pure loss of will."[1]

In the same oral history, Cernan responded to Schmitt's anger. "And don't forget, this was in 1972, so we had thirty years to go before the century was over. . . . I certainly felt that [we would return to the Moon] in the time frame of the twentieth century."[2]

How, in the three years after Apollo 11, had an achievement that had been greeted with near universal euphoria, become transformed into a topic of boredom and apathy?

From the perspective of the early 21st century, the enormity of the Apollo program's success may appear so removed from current daily reality that reflections about its accomplishment span a spectrum of emotions. Among those who witnessed it as it happened live on television, it is not uncommon to hear nostalgic longing for a time when America could confront a nearly impossible goal and work tirelessly and fearlessly to accomplish it. "What's become of the national can-do spirit that got us to the Moon?" is a question asked countless times in the decades since the 1970s, usually voiced in anger or sadness. Others, like historian Gerald DeGroot, who grew up during the days of Mercury, Gemini, and Apollo, warily look back on that time as an era when gullible American taxpayers were cynically manipulated by alarmist politicians and opportunistic promoters to sign off on a multi-billion dollar waste that accomplished virtually nothing for mankind.[3] For a smaller segment of the population born in the years since the Apollo landings, and whose upbringing and education were colored by a darker legacy of the 1970s—a pervasive distrust of conventional history and an awareness and suspicion of government deception—the Moon landings can only be explained by a realization that they are the result of a conspiratorial hoax.

To tell the story of Apollo and what followed in its wake is impossible without acknowledging and understanding various historic threads unique to its era. From the *Collier's* magazine editorial that introduced their space series in 1952, through the Sputnik scare of 1957, and leading to President Kennedy's May 1961 challenge to land a man on the Moon, the entire enterprise of manned space flight was framed by the news media and politicians using Cold War metaphors which warned of a clear and present danger to the security of the "free world." Over the years this context receded from immediacy and into memory as the civilian nature of NASA's structure and accomplishments justifiably defined its historic legacy. Yet, in 1969, the space race with the Soviet Union was still a substantial element of the story. Even on the eve of Apollo 11, there was speculation that the Russians might attempt to best the American mission with a surprise space spectacular. While Armstrong and Aldrin were on the Moon, the Soviet Union attempted to land an unmanned robot probe Luna 15 on the lunar surface. Launched three days before Apollo 11, it was sent on a mission to bring back the first lunar soil sample and steal some of the attention accorded the first manned landing. Alas, Luna 15 crash-landed just hours before *Eagle*'s liftoff, thus ensuring its propaganda status as little more than a historical footnote.

In 1969, before the eyes of the world, NASA met Kennedy's goal and left little doubt which country had won the space race. In response, the highly secretive Soviet program announced months later that their actual intended goal had been the establishment of orbiting space stations. Not surprisingly, reality was less definitive. Since the fall of the Soviet Union, a clearer history has emerged, revelaing that the Soviet manned lunar program encountered severe problems with its massive N1 Moon rocket, which exploded twice during unmanned launches in 1969. The program was definitively abandoned only in 1974.

Many who followed the Apollo 11 landing on television had also viewed President Kennedy's speech to Congress of May 25, 1961, eight years earlier. This address is now almost exclusively remembered for his challenge to land a man on the Moon. But the content of the address was largely dominated by economic, international, and defense concerns. Space was the ninth and final item, preceded by nearly an hour of lengthy reports on social progress abroad, self-defense, NATO, civil defense, Southeast Asia, and Latin America. The address was rich with the rhetoric of the Cold War and opened with warnings about the covert actions by "adversaries of freedom" (clearly understood as the USSR) and their recent activities interfering in conflicts in Asia, Latin America, Africa, and the Middle East "to exploit, to control, and finally to destroy the hopes of the world's newest nations." When Kennedy finally announced his audacious lunar goal, the geopolitics of the

1. Apollo 17 Lunar Surface Journal: EVA-3 Close-out; NASA, 1995. With annotations by Eric M. Jones. www.hq.nasa.gov/alsj/a17/a17.clsout3.html
2. Ibid.
3. Gerald DeGroot, *Dark Side of the Moon: The Magnificent Madness of the American Lunar Quest*. New York: NYU Press, 2006.

previous hour was fresh in the minds of the listeners. Though he persuasively made his case with a dash of idealism, the Moon landing was still placed squarely within the arena of the Cold War: "Now it is time to take longer strides—time for a great new American enterprise—time for this nation to take a clearly leading role in space achievement, which in many ways may hold the key to our future on Earth. For while we cannot guarantee that we shall one day be first, we can guarantee that any failure to make this effort will make us last."

There was more than geopolitics behind Kennedy's decision to undertake the challenge—and it was far more immediate and pragmatic than an attempt to motivate American idealism by exploiting the vision and dream of von Braun. It is not often remembered that, when President Kennedy delivered his May 1961 address, it came during the first months of an administration in crisis. The failed Bay of Pigs invasion of Cuba was a major international embarrassment for United States foreign policy and took place in April, just a week after Yuri Gagarin's historic manned orbital mission. Kennedy needed something to not only answer the Soviet triumph, but to demonstrate his administration's resolve and vision as well. As space historian Roger D. Launius described it, Kennedy's decision was the result of a "unique confluence of foreign policy crisis, political necessity, personal commitment and activism, scientific and technological ability, economic prosperity, and public mood."[4]

Even though Apollo was already on NASA's long-range plan when Gagarin was sent into orbit, his flight was but one element of many that instigated the Apollo program into reality. Again a surprise Soviet space first had forced an American president into action just as Sputnik led to increased funding for science education in the United States, NASA's formation, and the gradually reduced role of the military in the American space program.

The golden era of the American space program is often directly linked with the legacy of President Kennedy and the energy and idealism associated with his brief tenure in the Oval Office. Even though it is Richard Nixon's signature that appears on the lunar module plaques at the Apollo landing sites, every Moon flight departed from the Kennedy Space Center located on, what was then named, Cape Kennedy, in Florida. (Cape Canaveral was renamed Cape Kennedy in honor of the President following a recommendation by President Johnson six days after the assassination. The name was returned to Cape Canaveral in 1973, when the Florida Legislature passed a law restoring the former, 400-year-old, name.) Yet Kennedy's own personal feelings about the Apollo project and his motivations to undertake it were revealed when the Kennedy Library released a previously unknown audio recording in 2001. On November 21, 1962, President Kennedy met with NASA Administrator James Webb and other advisors. In the recording, Kennedy can be heard bluntly telling Webb, "Everything that we do should be tied into getting on to the Moon ahead of the Russians. We ought to get it really clear that the policy ought to be that this is the top priority program of the Agency and one of the top priorities of the United States government.... Otherwise we shouldn't be spending this kind of money, because I am not that interested in space. I think it's good. I think we ought to know about it. But we're talking about fantastic expenditures. We've wrecked our budget, and all these

President John F. Kennedy speaks to a gathering of media and employees at NASA Site 3 during a November 1962 visit. In his hands, he holds an early prototype model of the Apollo command module. Others seen are Vice President Lyndon B. Johnson; Dr. Robert R. Gilruth (at Johnson's left); and James E. Webb, NASA Administrator (at Kennedy's right).

4. Roger D. Launius, "Public opinion polls and perceptions of US human spaceflight," in *Space Policy* 19; 2003, pp. 163–175.

other domestic programs, and the only justification for it, in my opinion, is to do it in the time element I am asking."[5] In essence, Apollo was largely a high-priority international public relations gamble. And after an American footprint became the first to tread in lunar soil, those who understood this as the primary reason for its existence saw few reasons to return there.

Exclusive of Cold War politics, it is often assumed that there was widespread national enthusiasm about the space program during the Apollo era. Walter Cronkite's giddy "Oh boy!" on live television just after *Eagle* touched down, the seemingly myriad space-themed issues of *Life* that graced suburban coffee tables, and Tang commercials are iconic moments and images of the decade. However, as Launius has suggested, the belief that the American space program received wide public support and confidence during the 1960s is based largely on anecdotal accounts. Polling results from period indicate that the truth is more nuanced. Nearly all opinion polls conducted during the 1960s and early 1970s reveal that a majority of Americans thought the country was spending too much to explore space, and that the Apollo program was not worth the cost. The only poll that found a majority in favor was conducted in 1969, shortly after Apollo 11. (Launius also cites research that indicates widespread ignorance regarding the amount of taxpayer money expended on the space program; most people polled assumed space consumed a much larger percentage of the national budget than it actually did.)[6]

While the astounding accomplishment and human drama of the Apollo missions briefly captured the collective imaginations of the world's population, mass support in the U.S. was neither widespread nor very deep. And it was not uncommon during the early 1970s to encounter flippant comments that space program gave back little to the American public other than a box of rocks and Teflon. As Jack King, NASA Public Affairs officer at Kennedy Space Center during the Apollo years, recalled, "Once we had beaten the Russians, the American public started to focus their attention elsewhere, especially on the Vietnam War and the domestic front. There were so many things going on at the time. The interest starts to go down, and you have the three E's: energy, the economy, and the environment, and that's where the money started to go versus the space program."[7]

It's no surprise that national political trends and the concerns of the investment community keenly affected the health and fate of aerospace and military contractors of the era. Notably, after the early 1960s, marketing efforts by NASA contractors seldom overtly alluded to Cold War politics, reflecting in part NASA's and the government's desire for the Apollo program to be viewed as an apolitical technological and scientific quest. As public and political interest in the space program waned and dissatisfaction with the war in Vietnam increased, a number of these companies tried to diversify into different areas of business—from hotels to consumer electronics—in what became a period of growth for large, multi-industry conglomerates.

In NASA offices there was an awareness that times were changing. As the Vietnam War accelerated and took its toll on the national budget, President Johnson reduced NASA's expenditures, failing to approve major funding for post-Apollo projects. Shortly after the Moon landing, Wernher von Braun remarked in an interview with political scientist John Logsdon, "The legacy of Apollo has spoiled the people at NASA. They believe that we are entitled to this kind of a thing forever, which I gravely doubt. I believe that there may be too many people in NASA who at the moment are waiting for a miracle, just waiting for another man on a white horse to come and offer us another planet, like President Kennedy."[8]

Countless popular and scholarly histories have charted the massive social and cultural changes that the United States witnessed during the Apollo era. With the Vietnam War casting its large shadow on the late 1960s, there arose new awareness of social inequalities—racial, gender, and financial—and human values. Beliefs that largely went unquestioned at the beginning of the decade often gave rise to polarizing arguments ten years later that could range from the proper role of the United States in foreign affairs to humanity's relationship with the global environment.

Within this seismic cultural shift of values, attitudes toward the American space program were not always predictable. It was often simplistically assumed that Americans who traditionally supported their government in Vietnam, affixed flag decals to their cars, and affirmed conservative values also supported the country's space program. True to this stereotype, a 1971 episode of *All in the Family* features a moment

5. The audio recording of this meeting is available at the John K. Kennedy Presidential Library and Museum, and a full transcript of the meeting is available on NASA's history website. Reference Presidential recordings collection tape number 63.
6. Launius, op. cit.
7. Jack King interview with the authors, November 9, 2011.
8. T. A. Heppenheimer, "The Space Shuttle Decision," SP-4221, NASA History Office, 1998. Chapter 4.

(FX1)SAN FRANCISCO, Jan.30--LACK OF INTEREST IN MOONSHOOT--Author Norman
Mailer in San Francisco Friday said Americans are about as interested in
Sunday's Apollo 14 moonshoot "as a border war in Bolivia" and it's
a shame. The 47-year-old Pulitzer prize winner blamed the National Aeronautics and Space Administration for using the space program as a
"political football" instead of a means to bring the nation together.
(SEE STORY)(APWIREPHOTO)(ott70300stf-sjv) 1971

Author Norman Mailer at his San Francisco press conference bemoaning a lack of public interest in the pending launch of Apollo 14.

when working-class Archie Bunker proudly extols his ownership of a "genuine facsimile of the Apollo 14 insignia." In late 1968, NASA Director Thomas O. Paine called Apollo 8 "a triumph of the squares," attempting to contrast the virtues of hard-working, stoic, pocket protector-wearing engineers with the youthful, hirsute members of the counterculture variously depicted in news reports as experimenting with drugs and protesting on college campuses.[9]

America's most visible space enthusiast of the late 1960s was, of course, Walter Cronkite. "The most trusted man in America" (as he was once famously named in a widely reported opinion poll) was also critical of America's involvement in Vietnam. His February 1968 special report broadcast is often cited as a tipping point moment for swaying public opinion against the war. Yet his infectious enthusiasm for the space program—which a television news historian said made him sometimes seem "more cheerleader than reporter"[10]—didn't sway the opinion of his own 19-year-old daughter during the summer of 1969. "I was almost anti-NASA. . . . I thought we were just now bringing our own junk, golf balls, lunar module, to pollute the Moon," Nancy Cronkite told her father's biographer Douglas Brinkley.[11]

Associated with many liberal political and cultural causes, novelist Norman Mailer was a vocal critic of American foreign policy and an unsuccessful Democratic candidate in the New York mayoral primary of 1969. His reflections on the mission of Apollo 11 were serialized in *Life* magazine and later published as the book *Of a Fire on the Moon*. In early 1971, Mailer held a press conference in San Francisco where he railed against public apathy about the upcoming flight of Apollo 14, saying it was a shame that Americans were about as interested in the flight as "a border war in Bolivia." In particular he criticized NASA for using the space program as a "political football," rather than as a way to bring the nation together.[12] Seldom shy of the limelight or of making dramatic public pronouncements, Mailer notoriously did very little to bring the nation together three months after his Apollo 14 comments, when he appeared at New York's Town Hall with Germaine Greer and Diana Trilling to belligerently debate the subject of women's liberation.

Protests and Priorities

Public protest about class and racial divisions within the United States touched the Apollo program itself, when members of the Southern Christian Leadership Conference descended on Cape Kennedy to demonstrate at the launches of both Apollo 11 and Apollo 14. In July 1969, before Rev. Ralph Abernathy, who had succeeded Dr. Martin Luther King as the head of the SCLC, joined fellow SCLC leader Hosea Williams and twenty-five impoverished families in a literal "Poor People's Campaign" mule train to a location near Cape Kennedy, all three of the national television networks included reports about the demonstration as part of their extensive Apollo 11 live coverage. Williams was quoted in the press, explaining, "These demonstrations are not in protest of our ability to explore outer space, but in protest of Congress's inability to choose

9. Quoted in Robert Zimmerman's *Genesis: The Story of Apollo 8: The First Manned Flight to Another World*. New York: Four Walls Eight Windows, 1998, pg. 278.
10. Barbara Matusow, *Evening Stars: The Making of the Network News Anchor*. Boston: Houghton-Mifflin, 1983, p 127.
11. Douglas Brinkley, *Cronkite*. New York: HarperCollins, 2012, p. 414.
12. AP news report, January 30, 1971.

Members of the Southern Christian Leadership Conference stage a protest outside the Kennedy Space Center prior to the January 31, 1971, launch of Apollo 14.

priorities and bring to the nation the injustices and inequities of the space exploration appropriations as against appropriations to the poor."[13] Hoping to divert a public relations nightmare, NASA Administrator Paine and Public Relations head Julian Scheer met with Abernathy and Williams on the eve of the launch, at which time Paine was quoted as saying, "if we could solve the problems of poverty by not pushing the button to launch men to the Moon tomorrow, then we would not push that button." Paine and Scheer also offered bus transportation and exclusive viewing stand seating to 100 members of the Poor People's Campaign to witness the Saturn V launch the next day. Williams later admitted to an AP reporter, "I thought the launch was beautiful. The most magnificent thing I've seen in my whole life."[14]

In 1971, Abernathy and Williams returned with thirty black maids from Daytona Beach, who marched seventy-five miles to the Kennedy Space Center. The maids said they earned $35 a week cleaning $50-dollar-a-night motel rooms for the press corps and launch spectators. "Our country is spending $30 billion to bring men back from the Moon to get some Moon rocks for Vice President Spiro Agnew to hand out to heads of state," Williams said a press account. A fellow SCLC leader, Joseph Hammonds, got a bit more coverage with his assertion, "America is sending lazy white boys to the Moon because all they're doing is looking for Moon rocks. If there was work to be done, they'd send a nigger."[15]

An Icon of Unexpected Power

Race, gender, war, poverty, and cultural divides weren't the only issues vying for public attention during the later years of Apollo. An unpredictable confluence of events turned what many consider the Apollo's most iconic image into the symbol of a cultural concept and the spur for a philosophical movement. As historian Robert Poole has documented, before the first lunar flight, few at NASA had given much thought to the powerful impact the first photographs of Earth witnessed by human beings would have on the world. Presciently understanding their potential symbolic power, writer Stewart Brand began a campaign in 1966 to have NASA release the first photographs of the entire Earth seen from satellites. When Apollo 8 escaped Earth's gravitational influence and headed for the Moon, taking photographs of the Earth was not a major part of the flight plan. It was in a low-priority category labeled "targets of opportunity."[16] (In fact, a similar photographic oversight came to haunt Apollo 11, when it was discovered after its return that there were no good photographs of Neil Armstrong standing on the Moon; all of the famous images from that flight are of Aldrin taken by Armstrong.)

The moment the famous "Earthrise" photograph was taken, a transcript from the on-board audio recorder reveals the Apollo 8 crew excitedly scrambling to capture the image before it was lost. "Oh my God! Look at that picture over there! Here's the Earth coming up. Wow, is that pretty!" Bill Anders exclaims on the tape. As Anders, always by-the-book, takes a black-and-white photo of the scene, he is jokingly chided by Frank Borman, his commander: "Hey, don't take that, it's not scheduled!" On the recording can be heard laughter and then this priceless exchange, which captures the moment as one of the most famous photographs of the century is snapped:

ANDERS: You got color film, Jim? Hand me that roll of color quick, will you.

LOVELL: Oh man, that's great!

ANDERS: Hurry. Quick—

13. *Toledo Blade*, "Poor Heading For Apollo Site" (AP), July 14, 1969.
14. Hosea Williams citation.
15. *St. Petersburg Evening Independent*, Feb 1, 1971. (AP) "Poor People Protest 'Moon Rocks," p. 2.
16. Robert Poole, *Earthrise: How Man First Saw Earth*. New Haven: Yale University Press, 2008.

LOVELL: Take several of them! Here, give it to me.

ANDERS: Calm down, Lovell.

The color photograph, which has been poetically described as "an epiphany in space," was taken, in fact, by Anders. But it was also a group effort: Borman was turning the spacecraft to a new orientation, making it possible for the rising Earth to suddenly became visible in Anders's window. And, of course, it was Lovell who was able to find a color film magazine and give it to Anders for the iconic color image.[17]

As Stewart Brand realized two years earlier, such images, when shared with the rest of the world, would have a profound impact. Six years before the Apollo 8 photographs, Rachel Carson's *Silent Spring* was published, the moment often cited as the birth of the modern environmental movement. From it came the first legislation against pesticide spraying and increased awareness of what humans were doing to their ecosystem. A year before the Apollo 8 photographs were seen, the supertanker *Torrey Canyon* ran aground off the Western coast of England. At the time it was the worst oil spill in history, affecting hundreds of miles of coastline in the England, France, Guernsey, and Spain. In early 1969, while magazines featuring the Apollo 8 images on their covers were still displayed on newsstands, Americans read news of a massive blowout on an oil platform in Santa Barbara Channel. Within ten days, 80,000 to 100,000 barrels of crude oil spilled onto the beaches of in Southern California. Five months later came reports that the Cuyahoga River in Cleveland was so dangerously polluted it had spontaneously caught fire. These events surrounding the publication of the Earthrise photographs and the feelings they evoked, made many wonder why we were spending so much effort and money to examine the cold, dead, and barren surface of the Moon when our gaze might be better focused on our home planet and what we were doing to it.

When, during the spring of 1970, America observed the first Earth Day, NASA images of the Earth were ubiquitous. The same year also witnessed the signing of the National Environmental Policy Act and the formation of the Environmental Protection Agency.

As biologist Lewis Thomas wrote in 1974, "Viewed from the distance of the Moon, the astonishing thing about the Earth, catching the breath, is that it is alive." Thomas's thoughts were mirrored by former NASA researcher, chemist, and environmentalist James Lovelock, when his Gaia hypothesis—a theory that posits that the Earth is a self-regulating, living, complex, evolving system—gained public awareness during the 1970s.[18]

Philosophically turning our gaze Earthward, and looking at our home planet as a living organism, made Cold War geopolitics appear trivial. Writer and critic Jesse Kornbluth has suggested the fresh perspective that resulted from the Apollo 8 photographs "could be the most powerful thought of the twentieth century."[19].

NASA's evident effort to downplay any Cold War framing of America's quest for the Moon by the mid-1960s, suggests a conscious realization by its management that once the Moon had been reached by the United States, it would be difficult to justify funding of the magnitude it had been receiving. Even if the Soviet Union were to accomplish the goal first, setting up semi-permanent colonies on the Moon would be in the distant future. At the climatic moment during the summer of 1969, NASA's management attempted to define their lunar triumph as merely the first step in a cosmic saga of discovery and exploration that would play out during the waning years of the twentieth century. Speaking to reporters shortly before Apollo 11's July 16 launch, NASA Administrator Paine mused that, "the real goal is to develop and demonstrate the capability for interplanetary travel."[20] Paine was advocating for a $10 billion investment that would lead to lunar colonies, space stations, and a mission to Mars in 1983.

Paine's one ally in the White House was Vice President Spiro Agnew, chairman of the Space Task Group, a special panel created by President Nixon to determine NASA's future course after Apollo. The new Vice President made headlines on the eve of the Apollo 11 mission by openly calling for a manned mission to Mars. Even at what was the highpoint of public approval for the Apollo program, Agnew's flight of interplanetary fantasy was met with laughter and irritation. Already defined as a figure of ridicule by political humorists, Agnew was lampooned by one newspaper cartoonist as a clueless, grinning space cadet outfitted in an Apollo spacesuit holding a suitcase labeled, "Mars or Bust!" In the White House, where Agnew had little real persuasive influence, there was much less levity. Nixon was never enthusiastic about the exploration of space and his transition team had already

17. Poole, ibid. The abridged transcription of the on-board audio presented here is by Andrew Chaikin, who pointed out in a private communication with the authors the group effort that enabled the shots to be achieved.
18. Poole, ibid.
19. Jesse Kornbluth, review of *Earthrise*, in *Head Butler*, January 20, 2009.
20. Reginald Turnill, *The Moonlandings: An Eyewitness Account*, Cambridge, 2003. Quoted in DeGroot, op. cit.

Back to the Garden

In 1966, Stewart Brand, a veteran of Ken Kesey's Merry Pranksters and the organizer of the LSD-inspired Trips Festival, an event headlined by the Grateful Dead (and described in detail in Tom Wolfe's *Electric Kool-Aid Acid Test*), had another inspiration. "I herded my trembling thoughts together as the winds blew and time passed," he wrote. "A photograph would do it—a color photograph from space of the Earth. There it would be for all to see, the Earth complete, tiny, adrift, and no one would ever perceive things the same way." The next day, Brand ordered hundreds of buttons and posters that read: "Why haven't we seen a photograph of the whole Earth yet?" Wearing a white jump suit, boots, a top hat, and a day-glo sandwich board, Brand went to the gates of the University of California in Berkeley to sell his buttons. On November 10, 1967, NASA's ATS-3 satellite transmitted the first color photo of the whole Earth. Brand said the image "gave the sense that Earth's an island, surrounded by a lot of inhospitable space. And it's so graphic, this little blue, white, green and brown jewel-like icon amongst a quite featureless black vacuum."

The picture from ATS-3 and the pictures of the "Earthrise" from Apollo 8 became global phenomena, said by some to be the most reproduced pictures of all time, and a catalyst for a cultural realignment. Growing environmental concerns that percolated slowly in the 1950s in books such as Helen and Scott Nearing's *Living the Good Life* (1954), and gained strength with Rachel Carson's *Silent Spring* (1962), suddenly came into visible focus. In the fall of 1968, Brand published the first *Whole Earth Catalog*, with the ATS-3 photo on its cover. Its first words were "We are as gods and might as well get good at it." Brand's premise was that, given the necessary consciousness, information, and tools, people would reshape the world into something environmentally and socially sustainable.

These ideas would become a backdrop for Woodstock, in 1969, at which photos of the Earth in space would become thematic. That year, R. Buckminster Fuller, also part of Brand's circle, would publish his *Operating Manual for Spaceship Earth*, and MIT Press issued the architect Paolo Soleri's *Arcology: The City in the Image of Man*. The first Earth Day, planned at a UNESCO meeting in San Francisco in 1969, was held on April 22, 1970.

Brand was no Luddite. To the contrary, in San Francisco in 1968, he assisted the engineer Douglas Engelbart in what became known as "The Mother of All Demos," in which revolutionary computer technologies were demonstrated for the first time: e-mail, hypertext, the mouse, and video-conferencing— ideas that had reached currency in certain circles with the publication of Norbert Weiner's *Cybernetics*, in 1948. Brand would later become a mentor and friend of Steve Jobs. The idea of home computing was an intrinsic part of the *Whole Earth Catalog*; in 1970, Brand published the first issue of a do-it-yourself journal, *Radical Software*.

The energy that was once Apollo's was captured by a youthful public who turned their quest homeward, with a concern for planetary hygiene and social justice. Thanks to NASA, they saw the planet and it was theirs. It was the only spaceship they wanted for the moment.

Why haven't we
seen a
photograph
of the whole
Earth yet
?

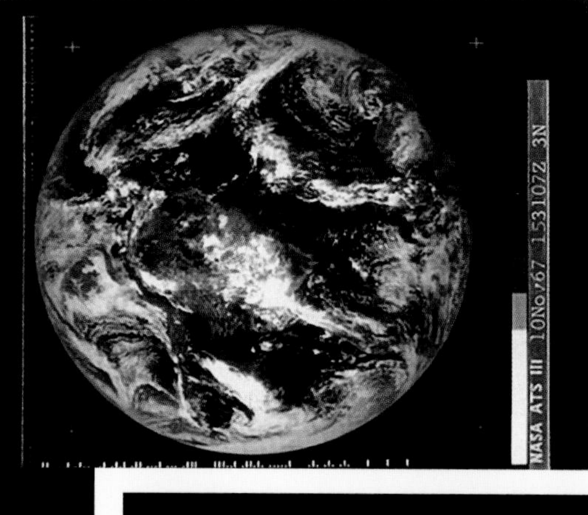

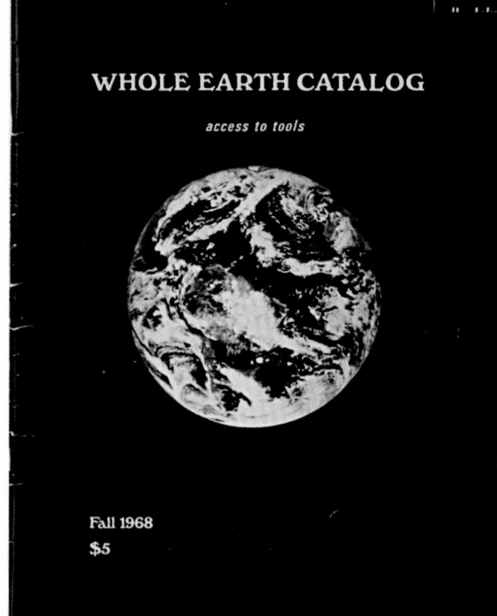

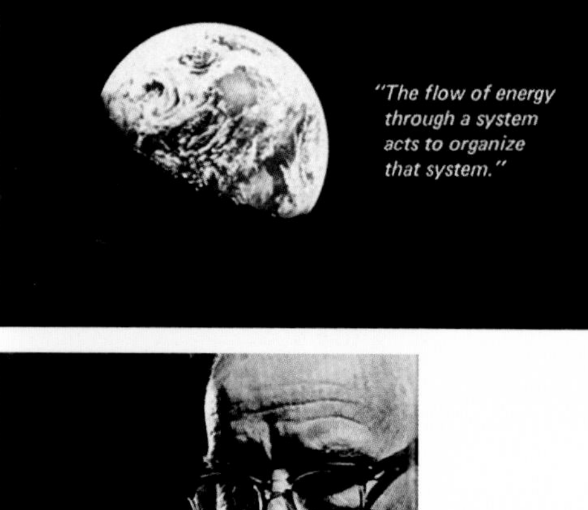

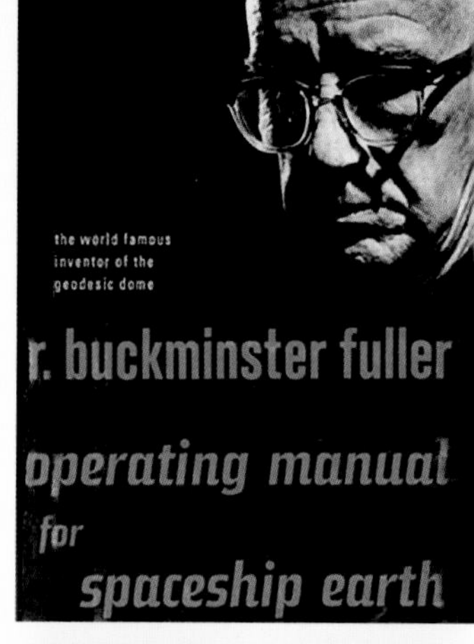

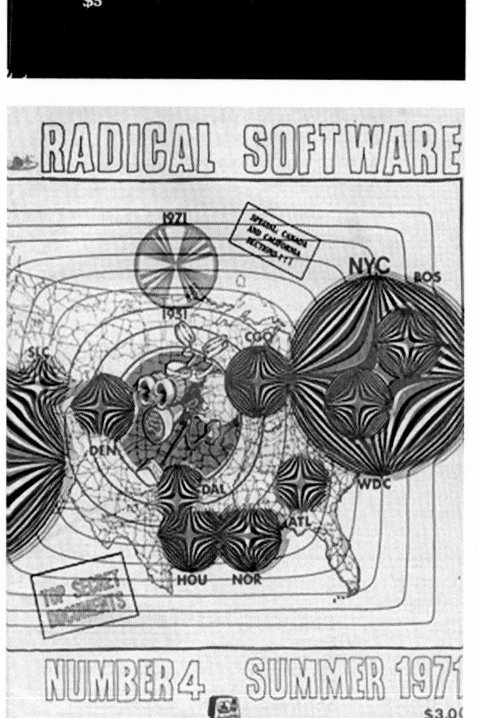

recommended avoiding making any ambitious plans beyond the Apollo missions, specifically a manned mission to Mars.

It fell to John Ehrlichman, Nixon's Assistant for Domestic Affairs, to rein in the rogue Vice President. "I was surprised by his obtuseness," Ehrlichman later recalled. "He had been [NASA's] guest of honor at space launchings, tours and dinners, and it seemed to me they had done a superb job of recruiting him to lead this fight to vastly expand their empire and budget." Ehrlichman described his message to Agnew as, "Look Mr. Vice President, we have to be practical. There is no money for a Mars trip. The president has decided that."[21]

On Capitol Hill, prominent democrats including the Senate Majority Leader Mike Mansfield, Senator Walter Mondale, and Congressman (and future New York City Mayor) Ed Koch all questioned the wisdom of a manned mission to Mars, as did Republican stalwart and member of the Senate Committee on Aeronautics and Space Sciences Margaret Chase Smith.[22] The final report of the Space Task Group reflected Nixon's earlier decision, and Agnew and the Nixon White House never raised the subject of Mars again.

While Vice President, Nixon had not only witnessed how the exploration of space could affect political fortunes, he saw how President Eisenhower used an American accomplishment in space as a public relations event with his recorded holiday greeting from the Project SCORE satellite. With Eisenhower's science advisor James Killian, Nixon advocated for the creation of a separate civilian space agency to carry out an open program of scientific activities, a decision that was to fundamentally affect what transpired during the decade that followed and how Apollo was seen by the rest of the world. In his memoirs, Nixon wrote, "Eisenhower finally came around to this view and approved a proposal for manned space vehicles. While he justified this decision on military grounds, I felt that something far more basic was involved. I believe that when a great nation drops out of the space race to explore the unknown, that nation ceases to be great."[23]

However, by the time he had reached the White House, Nixon did not have his eyes on the heavens. In his first inaugural address Nixon spoke of a nation "ragged in spirit . . . torn by division," and, at its conclusion, found an Earth-bound message from the flight of Apollo 8, which had taken place a few weeks earlier:

In that moment, their view from the Moon moved poet Archibald MacLeish to write:

To see the Earth as it truly is, small and blue and beautiful in that eternal silence where it floats, is to see ourselves as riders on the Earth together, brothers on that bright loveliness in the eternal cold — brothers who know now they are truly brothers.

In that moment of surpassing technological triumph, men turned their thoughts toward home and humanity — seeing in that far perspective that man's destiny on Earth is not divisible; telling us that however far we reach into the cosmos, our destiny lies not in the stars but on Earth itself, in our own hands, in our own hearts.

President Nixon shown in split-screen while sitting in the White House Oval Office speaking to Neil Armstrong and Buzz Aldrin while they walked on the lunar surface, July 20, 1969. Far from modest at this moment, Nixon informed the astronauts and the world in the second sentence of his brief call that "this certainly has to be the most historic telephone call ever made from the White House."

Six months later it was impossible not to look moonward again as the new President of the United States became a part of one of the most historic moments in human history. Despite grumblings among NASA officials about inserting time for a White House telephone call during the two-and-a-half hour EVA, Nixon succeeded in using the lunar landing to enhance his own image as President as more than a billion people watched live. Nixon fully embraced the powerful marketing impact of the lunar landing and took advantage of it in view of the whole world—on live television. This was an opportunity for the President to showcase the United States's technological superiority, not only in the lunar landing and exploration, but also in conducting the first Earth-to-Moon telephone call. Nixon successfully injected himself into the story.

But even while making history, Nixon couldn't escape the past. As many presidential historians have observed, the Kennedy legacy haunted the two men who filled the Oval

21. John Ehrichlman, *Witness to Power: The Nixon Years*. New York: Simon and Schuster, 1982, p. 145.
22. DeGroot, op. cit., pp. 246–247.
23. Richard Milhous Nixon, *RN: The Memoirs of Richard Nixon*. New York: Grosset & Dunlap, 1978, pp. 428–429.

President Nixon undertook a ten-day round-the-world trip in late July 1969, which began with a stop on the USS Hornet *to greet the Apollo 11 astronauts. Nixon's conversation with the astronauts, who were confined to their quarantine facility for three weeks, ranged from questions about the recent All-Star Game, whether any of them got sea sick during the recovery, and the President's memorable pronouncement: "This is the greatest week in the history of the world since the Creation."*

Office after the assassination. Nixon speech writer and *New York Times* columnist William Safire revealed that the White House staff underestimated the partisan public resentment against the newly elected President for putting his signature on the Apollo 11 commemorative plaque and phoning the astronauts on the Moon. "Presidents Kennedy and Johnson launched and encouraged the space program, grumped the *New York Times*, and it was 'unworthy' for President Nixon to 'share the stage' with the astronauts merely because he was in the White House 'by accident of the calendar' at its fruition," Safire recalled on the 20th anniversary of the Moon landing. "The *Washington Post* added that Nixon should not have signed the plaque because the Moon shot was no ordinary public works project." Safire noted that Nixon hoped "to use the American space triumph to override the public preoccupation with Vietnam."[24]

If there was any doubt in his mind that his enemies thought of Apollo as Kennedy's legacy, this reaction surely cemented it. Nixon greeted the crew of Apollo 11 on live television when they were brought aboard the USS *Hornet* and, a month later, presided over a large and lavish presidential dinner in their honor at the Century Plaza Hotel, in Los Angeles. In the audience were Vice President Spiro T. Agnew, Chief Justice Warren E. Burger, former Vice President Hubert Humphrey, fifty members of the U.S. House and Senate, forty-four governors, and fourteen Cabinet members. Later, in November, Nixon became the first sitting U.S. President to attend a space launch, when he witnessed the successful liftoff of Apollo 12.

The following year, while the nation watched the perilous return of Apollo 13, Nixon was personally briefed by two distinguished authorities who could speak with authority about the hazards involved: former Apollo 11 astronaut and Assistant Secretary of State for Public Affairs, Michael Collins, and Apollo 8 astronaut Bill Anders, who was then serving as Executive Secretary of the National Aeronautics and Space Council. Later, Nixon traveled to Hawaii, where he awarded astronauts Lovell, Haise, and Swigert the Presidential Medal of Freedom. "Greatness comes not simply in triumph but in adversity," Nixon said at the ceremony. "It has been said that adversity introduces a man to himself." Both the Apollo 11 and 13 missions had captivated the attention of the American people, and Nixon made sure he was there to celebrate their safe return. Subsequently, Nixon did little to instigate or reflect enthusiasm for the Apollo program. Returning crews visited the Oval Office and the White House issued official statements when appropriate. H.R. Haldeman's diary entry on the day of the Apollo 15 launch revealed Nixon's interest in space was mirroring the rapidly increasing disinterest elsewhere in the country: "The Apollo shot was this morning; the P [President Nixon] slept through it, but we put out an announcement that he had watched it with great interest."[25]

By this point in the Apollo story, Nixon and NASA had already been forced to cut the last three Apollo flights, which were originally to number twenty. In a March 7, 1970, announcement, Nixon stated, "We must think of [space activities] as part of a continuing process and not as a series of separate leaps, each requiring a massive concentration of energy. Space expenditures must take their proper place within a rigorous system of national priorities."

At a crucial moment during the summer of 1971, shortly after having slept through the 9:34 A.M. launch of Apollo 15, Nixon considered canceling the two remaining lunar missions, Apollo 16 and 17. Caspar Weinberger, Deputy Director of the Office of Management and Budget, opposed this proposal

24. William Safire, "Of Nixon, Kennedy and Shooting the Moon," *The New York Times*, July 17, 1989.
25. Alfred Robert Hogan, "Televising the Space Age: A Descriptive Chronology of CBS News Special Coverage of Space Exploration from 1957 to 2003," University of Maryland Masters Thesis, 2005.

The first sitting President of the United States to witness an American space launch in person, President Nixon sits in the viewing stand at Cape Kennedy for the November 14, 1969, blastoff of Apollo 12. The launch took place during a thunderstorm, during which the Saturn V was struck by lightning, nearly ending the mission during its first minute. NASA Administrator Thomas O. Paine (right) holds an umbrella over the President and First Lady, Pat Nixon. Daughter Tricia Nixon is seated in the row directly in front of the first couple.

arguing that necessary budget cuts should be exacted on programs that offer "no real hope for the future" rather than the Apollo program, which he believed gave "the American people a much needed lift in spirit (and the people of the world an equally needed look at American superiority)." In addition, Weinberger feared that, by eliminating Apollos 16 and 17, the President would be confirming a spreading belief that America's "best years are behind us."[26]

During the immediate post-Apollo 11 period at NASA, Paine and others were still making a concerted effort to think big and to make a case for Mars. An internal memo written in October 1969 reveals how NASA's new awareness of television's power to shape public opinion might aid in their long-term goals. A Gallup poll conducted that year found 53% of respondents opposed to funding a Mars program; only 39% were in favor.[27] To build enthusiasm for such a mission, a NASA Public Affairs administrator suggested that, during the upcoming Apollo 12 mission, time might be set aside during the moonwalk to include a live shot of Mars taken from the lunar surface while either Pete Conrad or Alan Bean would suggest, "This could be the next manned space flight goal of the eighties."[28]

Wounded by a series of rebuffs from the White House and by Nixon's April 1970 announcement that space expenditures should be placed within a system of national priorities, Paine made a final attempt to build support and construct ambitious plans by convening a three-day private retreat on the future of America's destiny in space. Held on Wallops Island off the coast of Virginia, where NASA operated a small rocket launch site to support science and exploration missions, the June 1970 conference brought together a room of visionaries and imaginative thinkers, including Wernher von Braun, Arthur C. Clarke, Robert Gilruth, George Mueller, and Neil Armstrong. Paine asked those in attendance to channel their "swashbuckling buccaneering courage" for "a completely uninhibited flow of new ideas." He called for discussions about new engines to extend human life out into the galaxy, an intercontinental space plane, global supercomputer networks, synthesized food manufacturing to free man from a 5000-year dependence on agriculture, and the "future evolution of terrestrial life to other worlds with eventual communication with other intelligence." He also advised the members of the Wallops Island retreat to be "competent and hard working, sensitive but steady nerved, visionary but tough minded, determined and thoughtful. . . . Be careful of ideology and amateur social science and economics."[29]

After the retreat, Paine wrote to Nixon: "The results are exciting and I would like to request an appointment to present to you our best current thinking. . . . The purpose is . . . to give you a heretofore unavailable Presidential-level, long-range view of man's future potential in space." According to historian Gerard J. DeGroot, the White House response was a polite "Thank you."

A liberal Democrat and a Johnson appointee, Paine understood his situation when dealing with a polarized White House consumed with a troubled national economy, war in Southeast Asia, budget cuts, domestic problems, and an upcoming reelection campaign. In late July of 1970, after serving as NASA Administrator for two-and-a-half years, during which he oversaw NASA's greatest triumph, Paine submitted his letter of resignation, choosing to return to his former employer, General Electric. By this time, von Braun had departed the Marshall Spaceflight Center to take on the role of NASA's Deputy Associate Administrator for Planning at NASA headquarters in Washington, where he witnessed the same budgetary wrangling and lack of enthusiasm for future goals that Paine encountered. Two years later, in May 1972, he also retired from NASA.

Meanwhile, NASA was moving ahead with planning for

26. "Memorandum For the President" by Caspar Weinberger (via George Shultz), August 12, 1971.
27. DeGroot, op. cit., p. 246.
28. Dwight Steven-Boniecki, *Live TV from the Moon*. Ontario: Apogee Books, 2010, p. 163.
29. Heppenheimer, op. cit., chapter 4.

Apollo 16 astronaut John Young jumps for joy and salutes as approval of the space shuttle budget is announced.

Skylab and the Space Shuttle, which was being promoted as a "space truck," a cheaper alternative to expendable rocket payload boosters. Against the advice of his many advisers who urged him to end the American human space program, Nixon approved the Shuttle's development in early 1972. Many assumed his decision was also political, wanting to demonstrably guarantee aerospace jobs in the electoral vote-rich state of California, home of North American Rockwell, the Shuttle's prime contractor. News of the House of Representative's approval of the 1973 space budget, which included the vote for the Space Shuttle, was reported to the Apollo 16 crew while they were walking on the Moon. When the news was relayed to John Young, who was later to commanded the first Shuttle mission, he responded, "The country needs that Shuttle mighty bad. You'll see."

The Press and the Later Apollo Flights

For the press, the later Apollo missions proved a challenge. The correspondents and reporters were acutely aware that reader and audience interest was diminishing rapidly. The networks severely reduced airtime, despite the far better live television images NASA was supplying. "Once we landed on the Moon and there was nothing there, it became a geology story," said Mark Bloom who covered the Apollo Moon missions for Reuters and the *New York Daily News*. "It was just a rocky, barren place, and a geology story isn't all that sexy compared with the adventure story of landing on the Moon for the first time."[30]

NASA's Public Affairs officers tried hard to sell stories about the scientific achievements of Apollo, and the technological innovations of contractor spin-offs. Lydia Dotto was one of the veteran reporters, who covered the later Apollo missions for Canada's *Globe and Mail*. "Try as they might, the science part of it wasn't really getting the PR job done," Dotto recalled. "Most people really didn't care if they brought back another bag of rocks from the Moon."[31]

"There was no way around it and NASA kind of knew it," agreed Bloom. "NASA pushed spin-offs very hard. I wrote a long article once for a medical publication that I worked for after Apollo was over, but the spin-offs were complete bullshit. All of that stuff was coming along anyway. They never got very far with it.

"The launches were still spectacular, though. I still couldn't get my friends interested. I'd tell them they should really go see a Saturn V liftoff. They'd say, 'Who cares?'"

The last Saturn V liftoff for a voyage to the Moon took place during the early hours of December 7, 1972. It was the first and only night launch of the world's most powerful rocket. John Logsdon was there and remembered it as "a thrilling experience," which was mixed with "a sense of melancholy that a great program, the Apollo trips to the Moon, were coming to an end. . . . There was no groundswell of public opinion saying let's continue flights to the Moon or start going to Mars. . . . The public was a little tired of the space program by the time of Apollo 17 and was ready to move on to other things."[33]

Another witness to the launch on that chilly December night was social philosopher, cultural critic, and poet William Irwin Thompson, who, alluding to the significance of the Apollo photos of the Earth in space wrote, "the recovery of our lost cosmic orientation will probably prove to be more historically significant than the design of the Saturn V rocket."[34]

30. Mark Bloom interview with the authors, January 13, 2012.
31. Lydia Dotto interview with the authors, January 20. 2012.
32. Bloom interview.
33. From "Is Another Moon Mission Written in the Stars?" NPR *Morning Edition*, December 7, 2012.

Gene Cernan, who rode Apollo 17 into the heavens that night, suggested that NASA might have been a victim of its own success. "It was the press who were saying, 'We've been to the Moon, so what are we going to do now? We've done that, and that was pretty easy. So, where to now, Columbus?'"[35]

With suppressed frustration, the man who has uneasily held the title "The Last Man on the Moon" for longer than he wished, spoke carefully about what America's reluctance to return to the Moon cost both the country and humanity. In an oral history recorded two decades after their mission, Cernan and Jack Schmitt reflected on the country's lost initiative. "The kids who are in grammar school now are the people who are going to be taking those trips back to the Moon and on to Mars. So we've got a generation in there that we've left in limbo. I don't really want to get into this," Cernan said, stopping himself, but then adding, "Quite frankly, I'm a little disappointed in us at this time, to know that we're really not much further along than we were back then." Apollo 17 lunar module pilot Schmitt interjected, "We're not as far. Then, at least we had a technology base. Now, we've got to rebuild it."[36]

This same worry was voiced in Caspar Weinberger's August 12, 1971, "Memorandum for the President," sent via George P. Shultz, then director of the Office of Management and Budget, as he made the case for keeping Apollo 16 and 17 in the budget: "It is very difficult to re-assemble the NASA teams should it be decided later, after major stoppages, to re-start some of the long-range programs."[37] For those like von Braun, who acutely understood the long-term implications of the decisions made in Washington in 1970 and 1971, these were signs that an era was rapidly coming to an end.

NASA's failure to articulate a clear vision about what was coming after Apollo was a failure of marketing. With four decades to reflect, many who were there at the time now look back on the missed opportunity. "We all thought the space program was beginning and it would last forever and just grow," recalled Alan Bean, Apollo 12 lunar module pilot and the fourth human to walk on the Moon. "So why would you worry about marketing when the tide of history is with you? People are going to keep going, we're going to build bases on the Moon, we're going to go to Mars. Just figure out how to do it and then we'll go do it. We don't need to convince anybody to like it, because we liked it."[38]

The "Blue Marble" was the name given to a series of whole earth photographs, including this one, taken by the crew of Apollo 17 on December 7, 1972.

Just like a successful consumer product, there's no guarantee that success will last forever. In fact, reversing the inevitable downturn in a product's life cycle is one of the most difficult jobs marketers face. As with countless product marketers who have failed to revitalize a fading brand, NASA wasn't able overcome the declining interest after the tremendous success of Apollo 11.

The advances in American technology achieved since the early 1970s have defined the half-century since the end of Apollo. While further research and development into large payload vehicles has not progressed much beyond the Saturn V, the startling contrast between the Raytheon-manufactured

34. Poole, op. cit., introduction.
35. Gene Cernan interview with the authors, February 24, 2012.
36. Apollo 17 Lunar Surface Journal, NASA, 1995.
37. Caspar Weinberger, Memorandum for the President, op. cit.
38. Alan Bean, interview with the authors, November 20, 2011.

The "Death Star" from one of George Lucas's Star Wars *films, looking very moonlike.*

Apollo Guidance Computer of the command module, perhaps the most sophisticated portable computer of its day, and handheld smart phones of the twenty-first century gives one pause. The Apollo Guidance Computer had a 2k memory with read-only storage of 32k. The earliest desktop computers marketed to consumers of the late 1970s were eight times more powerful than Apollo's navigational brain. A popular comparison of the size of the memory contained in a smartphone manufactured a half-century after the Apollo Guidance Computer compares the thickness of a single deck of playing cards to the height of the Empire State Building.

The generation of American visionaries who made real the digital revolution that defines the early 21st century were teenagers when Armstrong set foot on the Moon. They benefited from the increased emphasis on science and mathematics in schools during the years immediately after the launch of *Sputnik 1*.[39] In their secondary schools, many had access to computer terminals tied in to giant university mainframes that, when not running programming in support of projects related to the American space program, made off-time available to high school computer hobbyists. Witnessing the concerted effort to land a man on the Moon as they grew up, they understood that they were living in a time of rapid change. Many read books such as Arthur C. Clarke's *Profiles of the Future,* first published in 1962, which envisioned among many marvels an easily accessible global library containing the all world's information. (Clarke's global library and supercomputer networks were among the subjects he and other futurists discussed at NASA's retreat on Wallops Island in June 1970.) In 1962, Clarke forecast the global library for 2005. The actual work on something close to it began more than a decade earlier.

39. Paul Dickson, *Sputnik: The Shock of the Century*. New York: Walker & Company, 2001, p. 201.

The generation of Steve Jobs, Steve Wozniak, Bill Gates, and the hundreds of thousands of others who changed how we live, play, and think came of age during the heyday of the American space program, with its attendant increased funding for education, science, and technology. In essence, the visionaries of the digital generation were also the Apollo generation.

During the immediate post-Apollo years, the three manned Skylab missions (1973–1974) and the Apollo-Soyuz project of 1975 generated little public interest. Once again, television coverage was minimal, even during the initial days of the first manned Skylab flight, when the crew performed spacewalks to repair the vessel itself, previously damaged during deployment and in danger of overheating.

The next huge surge of American interest in space came during the summer of 1977. A film for which many at 20th Century Fox had confessed modest expectations became a unique pop-culture phenomenon that eventually altered the course of the American movie industry for future decades. *Star Wars* was an imaginative return to the pure adventure and kitsch of the Flash Gordon serials of the 1930s, albeit freshened by the latest special effects and the music of John Williams. In George Lucas's universe, the longuers of weightless travel and collecting rock samples were far from anyone's concern. As audiences around the globe followed the adventures of Luke Skywalker in a galaxy far, far away, they once again discovered the adventure, romance, and thrill that interplanetary travel promised in the narratives of Jules Verne and his science fiction descendants. This was the realm of outer space they chose to return to again and again in the decades to come. Eventually, this adventure became part of everyday life, carried in your pocket and delivered into the palm of your hand. ⊙

The Privatization of Space

In 1865, Jules Verne's visionary hero, Impey Barbicane, financed his trip to the Moon with international subscriptions from donors expecting not the slightest chance of profit. While most people in the 21st century would find it difficult to name a single contemporary NASA astronaut or administrator, less challenging would be the task of identifying Barbicane's modern counterparts in the private sector. Among the high-profile personalities who have captured the public's imagination and ignited a new enthusiasm for humankind's next steps into the cosmos: SpaceX entrepreneur Elon Musk (top left), whose privately funded space transportation company has carried cargo to the International Space Station; Austrian skydiver Felix Baumgartner (top right) departing the Red Bull Stratos capsule twenty-four miles above the Earth in his historic October 2012 free-fall space dive; and Sir Richard Branson (bottom left) whose space tourism company, Virgin Galactic, plans to take paying passengers into suborbital space on vehicles like the VSS Enterprise (bottom right), pictured during a 2010 test flight.

Selected Bibliography

PRINT RESOURCES

Agel, Jerome, ed. *The Making of Kubrick's 2001*. New York: New American Library, 1970

Aldrin, Buzz, and John Barnes. *The Return*. New York: Tom Dehetry, 2001.

Aldrin, Buzz, and Ken Abraham. *Magnificent Desolation: The Long Journey Home from the Moon*. New York: Harmony, 2009.

Aldrin, Buzz, and Malcom McConnell. *Men From Earth*. New York: Bantam, 1989.

Aldrin, Edwin E. "Buzz" Jr. and Wayne Warga. *Return to Earth*. New York: Random House, 1973.

Allen, Michael. *Live From the Moon: Film, Television and the Space Race*. London: I. B. Tauris, 2009.

Analysis of Apollo 8: Photography and Visual Observations. Washington, D.C.: NASA, 1969. SP-201.

Apollo 11 Spacecraft Commentary, July 16–24, 1969. NASA Manned Spacecraft Center, Houston.

Armstrong, Neil, Michael Collins, and Edwin E. Aldrin, Jr. *First on the Moon*. Boston: Little, Brown, 1970.

Arnold, H.J.P., ed. *Man in Space: An illustrated History of Spaceflight*. New York: Smithmark, 1993.

Balker, David. *The History of Manned Space Flight*. New York: Crown, 1981.

Barbour, John, and the Writers and Editors of The Associated Press. *Footprints on the Moon*. New York: Associated Press, 1969.

Barbree, Jay. *Live from Cape Canaveral: Covering the Space Race, from Sputnik to Today*. New York: Smithsonian Books/Collins, 2007.

Bassior, Jean-Noel. *Space Patrol: Missions of Daring in the Name of Early Television*. Jefferson, N.C.: McFarland & Company, 2005.

Bean, Alan. *Painting Apollo: First Artist on Another World*. Washington, D.C.: Smithsonian Books, 2009.

Bean, Alan, and Andrew Chaikin. *Apollo: An Eyewitness Account by Astronaut/Explorer/Artist/Moonwalker*. Shelton, Conn.: Greenwich Workshop Press, 1998.

Benford, Gregory. T*he Wonderful Future That Never Was*. New York: Hearst Books, 2010.

Benson, Charles D., and William B. Faherty. *Gateway to the Moon: Building the Kennedy Space Center Launch Complex*. Gainesville: University Press of Florida, 1978.

Benson, Charles D., and William Barnaby Faherty. *Moon Launch! A History of the Saturn-Apollo Launch Operations*. Gainesville: University Press of Florida, 2001.

Benson, Charles D., and William Barnaby Faherty. *Moonport: A History of Apollo Launch Facilities and Operations*. Washington, D.C.: NASA, 1978. SP-4204.

Biddle, Wayne. *Dark Side of the Moon: Wernher von Braun, the Third Reich, and the Space Race*. New York: W.W. Norton, 2009.

Bilstein, Roger E. *Stages to Saturn: A Technological History of the Apollo/Saturn*. Washington, D.C.: NASA, 1980. SP-4206.

Black, Conrad. *Richard M. Nixon: A Life in Full*. Public Affairs, 2007.

Blair, Don. *Splashdown! NASA and the Navy*. Nashville: Turner, 2004.

Bonestell, Chesley, and Willy Ley. *The Conquest of Space*. New York: Viking, 1949

Booker, M. Keith. *Science Fiction Television*. Westport, Conn.: Praeger, 2004.

Boomhower, Ray E. *Gus Grissom: The Lost Astronaut*. Indianapolis: Indiana Historical Society Press, 2004.

Borman, Frank and Robert J. Serling. *Countdown: An Autobiography*. New York: Silver Arrow, 1988.

Braudy, Leo. *The Frenzy of Renown: Fame and Its History*. New York: Oxford University Press. 1986.

Brinkley, David. *David Brinkley: A Memoir*. New York: Knopf, 1995.

Brinkley, Douglas. *Cronkite*. New York: HarperCollins, 2012.

Brooks, Courtney G., James M. Grimwood, and Loyd S. Swenson Jr. *Chariots for Apollo: A History of Manned Lunar Spacecraft*. Washington, D.C.: NASA, 1979. SP-4205.

Buckbee, Ed with Schirra, Wally. *The Real Space Cowboys*. Burlington, Ont.: Apogee, 2005.

Burrough, Bryan. *Dragonfly: NASA and the Crisis Aboard Mir*. New York: Harper Collins, 1998.

Burrows, Williams E. *This New Ocean: The Story of the First Space Age*. New York: Random House, 1998.

Byrnes, Mark E. *Politics and Space: Image Making by NASA*. Westport, Conn.: Praeger, 1994.

Carpenter, M. Scott, L. Gordon Cooper Jr., John H. Glenn Jr., Virgil I. Grissom, Walter M. Schirra Jr., Alan B. Shepard Jr., and Donald K. Slayton. *We Seven*. New York: Simon & Schuster, 1962.

CBS News and CBS Television Network. *10:56:20 PM EDT, 7/20/69: The Historic Conquest of the Moon as Reported to the American People*. New York: CBS, 1970.

Cernan, Eugene, and Don Davis. *The Last Man on the Moon: Astronaut Eugene Cernan and America's Race in Space*. New York: St. Martin's, 1999.

Chaikin, Andrew. *A Man on the Moon: The Voyages of the Apollo Astronauts*. New York: Viking, 1994.

Chaikin, Andrew, and Alan Bean. *Mission Control, this is Apollo: The Story of the First Voyages to the Moon*. New York: Viking, 2009.

Chaikin, Andrew, and Victoria Kohl. *Voices from the Moon: Apollo Astronauts describe their Lunar Experiences*. New York: Viking, 2009.

Cirino, Robert. *Power to Persuade: Mass Media and the News*. New York: Bantam, 1974.

Clarke, Arthur C. *Profiles of the Future*. New York: Harper & Row. 1962.

Collins, Michael. *Carrying the Fire: An Astronaut's Journeys*. New York: Farrar, Straus and Giroux, 1974.

Compton, William David. *Where No Man has Gone Before: A History of Apollo Lunar Exploration Missions*. Washington, D.C.: NASA, 1989. SP-4214.

Conrad, Nancy, and Howard A. Klausner. Rocket Man: *Astronaut Pete Conrad's Incredible Ride to the Moon and Beyond*. New York: New American Library, 2005.

Cooper, Jr., Henry S. F. *Apollo on the Moon*. New York: Dial, 1969.

Cooper, Jr., Henry S. F. *Moon Rocks*. New York: Dial, 1970.

Cooper, Jr., Henry S. F. *13: The Flight That Failed*. New York: Dial, 1973.

Cortright, Edgar M. ed. *Exploring Space with a Camera*. Washington, D.C.: NASA, 1968. SP-168.

Cortright, Edgar M., ed. *Apollo Expeditions to the Moon*. Washingto,n D.C.: NASA, 1975.

Cronkite, Walter. *A Reporter's Life*. New York: Knopf, 1996.

Cronkite, Walter, and Don Carleton. *Conversations with Cronkite*. Austin: University of Texas Press. 2010.

Cunningham, Walter. *The All-American Boys*. New York: Macmillan, 1977.

D'Antonio, Michael. *A Ball, a Dog, and a Monkey: 1957 — The Space Race Begins*. New York: Simon & Schuster. 2007.

Darling, David. *The Complete Book of Spaceflight: From Apollo 1 to Zero Gravity*. Hoboken, N.J.: Wiley, 2003.

Dasch, E. Julius, ed. *A Dictionary of Space Exploration*. New York: Oxford University Press, 2005.

Dawson, Virginia D., and Mark D. Bowles, eds. *Realizing the Dream of Flight*. Washington D.C.: NASA. 2005. SP-4112.

Day, James. *The Vanishing Vision: The Inside Story of Public Television*. Berkeley: University of California Press, 1995.

DeGroot, Gerard J. *Dark Side of the Moon: The Magnificent Madness of the American Lunar Quest*. New York: NYU Press, 2006.

Dethloff, Henry C. *Suddenly, Tomorrow Came . . . A History of the Johnson Space Center*. Washington D.C.: NASA, 1993. SP-4307.

Dick, Steven J., ed. *NASA's First 50 Years: Historical Perspectives*. NASA SP-2010-4704.

Dickson, Paul. *Sputnik: The Shock of the Century*. New York: Walker, 2001.

Donovan, Robert J., and Ray Scherer. *Unsilent Revolution: Television News and American Public Life, 1948–1991*. Cambridge, UK: Cambridge University Press, 1992.

Duin, Steve, and Mike Richardson, and S. Mark Young. *Blast Off! Rockets, Robots, Ray Guns, and Rarities from The Gold Age of Space Toys*. Milwaukie, Ore.: Dark Horse, 2001.

Duke, Charlie, and Dotty Duke. *Moonwalker*. Nashville: Oliver-Nelson, 1990.

Earth Photographs from Gemini VI through XII. Washington, D.C.: NASA, 1968. SP-171.

Ehrichlman, John. *Witness to Power: The Nixon Years*. New York: Simon & Schuster, 1982.

Ertel, Ivan D., and Roland W. Newkirk, with Courtney G. Brooks. *The Apollo Spacecraft: A Chronology*. 4 vols. Washington, D.C.: NASA, 1969. SP-4009.

Ezell, Edward Clinton, and Linda Neuman Ezell. *The Partnership: A History of the Apollo-Soyuz Test Project*. Washington, D.C.: NASA 1978. SP-4209.

French, Francis, and Colin Burgess. *Into that Silent Sea: Trailblazers of the Space Era, 1961–1965*. Lincoln: University of Nebraska Press, 2007.

Froehlich, Walter. *Apollo Soyuz*. Washington, D.C.: NASA, 1976.

Glenn, John. *Letters to John Glenn*. Houston, Tex.: World Book Encyclopedia Science Service, 1964.

Glenn, John, and Nick Taylor. *John Glenn: a Memoir*. New York: Bantam, 1999.

Godbold, James M. *All Aboard: Lucky in War, Lucky in Peace, Lucky in Love*. New York: iUniverse, 2003.

Gordon, Theodore J., and Julian Scheer. *First into Outer Space*. New York: St. Martin's, 1959.

Hacker, Barton C., and James M. Grimwood. *On the Shoulders of Titans: A History of Project Gemini*. Washington D.C.: NASA, 1977. SP-4203

Hansen, James R. *First Man: The Life of Neil A. Armstrong*. New York: Simon & Schuster, 2005.

Hartland, David M. *The First Men on the Moon: The Story of Apollo 11*. New York: Springer, 2007.

Heppenheimer, T. A. *The Space Shuttle Decision*. Washington D.C.: NASA History Office, 1998. SP-4221.

Hewitt, Don. *Tell Me A Story: Fifty Years and 60 Minutes in Television*. Public Affairs, 2002.

Hogan, Robert Alfred. "Televising The Space Age: A Descriptive Chronology of CBS News Special Coverage of Space Exploration From 1957 to 2003." Masters Thesis: University of Maryland, 2005.

Hurt, Harry III. *For All Mankind*. New York: Morgan Entrekin, 1988.

Irwin, James B., and William A. Emerson, Jr. *To Rule the Night: The Discovery Voyage of Astronaut Jim Irwin*. New York: A.J. Holman, 1973.

Johnston, Lyle. *"Good Night, Chet": A Biography of Chet Huntley*. Jefferson, N.C.: McFarland, 2003.

Johnson, Stephen B. *The Secret of Apollo: Systems Management in American and European Space Programs*. Baltimore: Johns Hopkins University Press, 2002.

Kauffman, James L. *Selling Outer Space: Kennedy, the Media, and Funding for Project Apollo, 1961–1963*. Tuscaloosa: University of Alabama Press, 1994.

Klerkx, Greg. *Lost in Space: The Fall of NASA and the Dream of a New Space Age*. New York: Pantheon, 2004.

Kraft, Christopher C. *Flight: My Life in Mission Control*. New York: Dutton, 2001.

Kranz, Gene. *Failure Is Not an Option: Mission Control from Mercury to Apollo 13 and Beyond*. New York: Simon & Schuster, 2000.

Lambright, W. Henry. *Powering Apollo: James E. Webb of NASA*. Baltimore: Johns Hopkins University Press, 1995.

Lattimer, Dick. *All We Did Was Fly To The Moon*. Gainesville: The Whispering Eagle Press, 1985.

Launius, Roger D. "Public Opinion Polls and Perceptions of US Human Spaceflight," in *Space Policy* 19, 2003.

Launius, Roger D., and Howard E. McCurdy. *Imagining Space: Achievements, Predictions, Possibilities, 1950–2050*. San Francisco: Chronicle Books, 2001.

Launius, Roger D., and Bertram Ulrich. *NASA and the Exploration of Space*. New York: Stewart, Tabori & Chang, 1998.

Levine, Arnold S. *Managing NASA in the Apollo Era*. Washington D.C.: NASA, 1982. SP-4102.

Ley, Willy, and Wernher von Braun. *The Exploration of Mars*. New York: Viking, 1956.

Lindsay, Hamish. *Tracking Apollo to the Moon*. London: Springer, 2001.

Logsdon, John M. *The Decision to Go to the Moon: Project Apollo and the National Interest*. Cambridge, Mass.: MIT Press, 1970.

Lovell, Jim, and Jeffrey Kluger. *Lost Moon: The Perilous Voyage of Apollo 13*. Boston: Houghton-Mifflin, 1994.

MacKinnon, Douglas, and Joseph Baldanza. *Footprints: The 12 Men who Walked on the Moon Reflect on their Flights, their Lives, and the Future*. Washington, D.C.: Acropolis, 1989.

Mailer, Norman. *Of a Fire on the Moon: A Work in Three Parts*. Boston: Little, Brown, 1970.

Makemson, Harlen. *Media, NASA, and America's Quest for the Moon. New York*: New York: Peter Lang, 2009.

Matusow, Barbara. *Evening Stars*. Boston: Houghton-Mifflin, 1983.

McAleer, Neil. *Odyssey: The Authorized Biography of Arthur C. Clarke*. London: Victor Gollancz. 1992.

McCurdy, Howard E. *Space and the American Imagination*. Baltimore: John Hopkins University Press, 2011.

Mersch, C. L. *The Apostles of Apollo: The Journey of the Bible to the Moon and the Untold Stories of America's Race into Space*. Bloomington, Ind.: iUniverse, 2010.

Miller, Ron, and Frederick C. Durant III. *The Art of Chesley Bonestell*. London: Paper Tiger, 2001.

Mitchell, Edgar, and Dwight Williams. *The Way of the Explorer: An Apollo Astronaut's Journey through the Material and Mystical Worlds*. New York: Putnam, 1996.

Murray, Charles A., and Catherine Bly Cox. *Apollo*. New York: Simon & Schuster, 1989.

Murray, Michael D., ed. *Encyclopedia of Television News*. Westport, Conn.: Greenwood, 1998.

Nadel, Norman. "Bil Baird Conquers Earth and *Space," in Columbus Citizen-Journal*, November 21, 1969.

Neufeld, Michael J. *Von Braun: Dreamer of Space, Engineer of War*. New York: A.A. Knopf, 2007.

Nixon, Richard Milhous. *RN: The Memoirs of Richard Nixon*. New York: Gosset & Dunlap, 1978.

Ordway, Frederick Ira. *Visions of Spaceflighy: Images from the Ordway Collection*. New York: Four Walls Eight Windows, 2001.

Ordway, Frederick Ira, and Randy Liebermann, eds. *Blueprint for Space: Science Fiction to Science Fact*. Washington, D.C.: Smithsonian Institution Press, 1992.

Orloff, Richard W. *Apollo by the Numbers: A Statistical Reference*. Washington D.C.: NASA, 2000. SP-2000-4029

Pellegrino, Charles R., and Joshua Stoff. *Chariots for Apollo: The Making of the Lunar Module*. New York: Atheneum, 1985.

Pogue, William R. *But for the Grace of God: An Autobiography of an Aviator and Astronaut*. Rorgers, Ark.: Soar with Eagles, 2011.

Poole, Robert. *Earthrise: How Man First Saw the Earth*. New Haven: Yale University Press, 2008.

Powell-Willhite, Irene E., ed. *The Voice of Dr. Wernher von Braun*. Burlington, Ont.: Apogee, 2007.

Prelinger, Megan. *Another Science Fiction: Advertising the Space Race 1957-1962*. New York: Blast Books, 2010.
Reed, Walt. *The Illustrator in America, 1860–2000*. New York: Watson-Guptill, 2001.
Ryan, Cornelius, ed. *Across the Space Frontier*. New York: Viking, 1952.
Ryan, Cornelius, ed. *Conquest of the Moon*. New York: Viking, 1953.
Salo, Edward George. "'Some People Call Me a Space Cowboy': The Image of the Astronaut in Life Magazine 1959–1972." Masters Thesis, Middle Tennessee State University, 1998.
Schefter, James. *The Race: The Uncensored Story of How America Beat Russia to the Moon*. New York: Doubleday, 1999.
Schirra, Wally, and Richard N. Billings. *Schirra's Space*. Boston: Quinlan Press, 1988.
Schmitt, Harrison H. *Return to the Moon: Exploration, Enterprise, and Energy in Human Settlement of Space*. New York: Praxis, 2006.
Scott, David, and Alexei Leonov. *Two Sides of the Moon*. New York: Thomas Dunne, 2004.
Shayler, David J. *Disasters and Accidents in Manned Spaceflight*. Chichester, UK: Praxis, 2000.
Shepard, Alan B., and Donald K. Slayton. *Moon Shot: The Inside Story of America's Race to the Moon*. Atlanta: Turner, 1994.
Siner, Charles K., Jr., "You and Neil and Buzz—You Made it" in *The Quill*, vol. 57, no. 9, September 1969.
Slayton, Donald K., and Michael Cassutt. *Deke!: U.S. Manned Space, From Mercury to the Shuttle*. New York: Forge, 1994.
Smith, Andrew. *Moon Dust: In Search of the Men Who Fell to Earth*. London: Bloomsbury, 2005.
Stafford, Thomas P., and Michael Cassutt. *We Have Capture: Tom Stafford and the Space Race*. Washington: Smithsonian Institution Press, 2002.
Star, Kristen Amanda. "NASA's Hidden Power: NACA/NASA Public Relations and the Cold War, 1945–1967." Ph.D. dissertation, Auburn University, 2008.
Steinberg, Cobbett. *TV Facts*. New York: Facts on File, 1985.
Steven-Boniecki, Dwight. *Live TV from the Moon*. Burlington, Ont.: Apogee, 2010.
Stoff, Joshua. *Building Moonships: The Grumman Lunar Module*. Charleston, S.C.: Arcadia, 2004.
Stuckey, Mary E. *Slipping the Surly Bongs: Reagan's Challenger Address*. College Station: Texas A&M University Press, 2006.
Swanson, Glen E., ed. *"Before This Decade is Out...": Personal Reflections of the Apollo Program*. Washington D.C.: NASA, 1999. SP-4223.
Swenson, Loyd S., James M. Grimwood, and Charles C. Alexander. *This New Ocean: A History of Project Mercury*. Washington D.C.: NASA, 1966. SP-4201.
Thompson, Neal. *Light This Candle: The Life and Times of Alan Shepard, America's First Spaceman*. New York: Crown, 2004.
The Today Show Looks at Ten Years of Space Exploration. Washington, D.C.: Aerospace Industries Association of America, 1968.
Tyson, Neil deGrasse. *Space Chroncles: Facing the Ultimate Frontier*. New York: W.W. Norton, 2012.
Ward, Bob. *A Funny Thing Happened On the Way To The Moon*. Greenwich, Conn.: Fawcett, 1969.
Ward, Bob. *Dr. Space: The Life of Wernher von Braun*. Annapolis, Md.: Naval Institute Press, 2005.
Watkins, Billy. *Apollo Moon Missions: The Unsung Heroes*. Westport, Conn.: Praeger, 2006.
Whitehouse, David. *One Small Step: The Inside Story of Space Exploration*. London: Quercus, 2009.
Woods, David W. *How Apollo Flew to the Moon*. New York: Springer, 2008.
Worden, Al, and Francis French. *Falling to Earth: An Apollo 15 Astronaut's Journey to the Moon*. Washington, D.C.: Smithsonian Books, 2011.
Worden, Alfred M. *Hello Earth: Greetings From Endeavour*. Los Angeles: Nash, 1974.
Von Braun, Wernher, and Frederick I. Ordway III. *History of Rocketry and Space Travel*. New York: Crowell, 1966.
Von Braun, Wernher, Fred L. Whipple, and Willy Ley. *Conquest of the Moon*. New York: Viking, 1953.
Zimmerman, Robert. *Genesis: The Story of Apollo 8: The First Manned Flight to Another World*. New York: Four Walls Eight Windows, 1998.
Zimmerman, Robert. *Leaving Earth: Space Stations, Rival Superpowers, and the Quest for Interplanetary Travel*. Washington, DC: Joseph Henry, 2003.

DVD

Apollo 11: A Night to Remember. Acorn Media., 2009
For All Mankind. The Criterion Collection, 2009.
In the Shadow of the Moon. THINKFilm and Lionsgate, 2007.
Live from the Moon: The Story of Apollo Television. Spacecraft Films, 2009.
Man on the Moon. CBS, 2003.
Manned Spacecraft Center Reports, 1964–1969. Spacecraft Films, 2007.
Moon Beat: A Documentary Film. NCI, 2009.
The Wonder of it All: Real American Heroes. Indican Pictures, 2009.
Tomorrowland: Disney in Space & Beyond. Disney, 2004.

PRIMARY INTERVIEWS

Bean, Alan (2011, November 20), NASA astronaut (Apollo 12, Skylab 3): telephone interview.
Biggs, Charles (2011, November 10), NASA Public Affairs: telephone interview.
Blair, Donald (2012, January 18), Mutual Radio Network: telephone interview.
Bloom, Mark (2012, January 13), Reuters & NY Daily News: telephone interview.
Buckbee, Edward (2012, January 10), NASA Public Affairs: telephone interview.
Button, Bob (2012, March 13), NASA Public Affairs, Grumman PR, and TRW PR: telephone interview.
Carey, Timothy (2013, March 13), Vice President, Raytheon
Carr, Harold (2011, December 14), Boeing Public Relations: telephone interview.
Cernan, Eugene A. (2012, February 24), NASA Astronaut (Gemini 9A, Apollo 10, Apollo 17): telephone interview.
Chudwin, David (2012, March 30), Apollo 11 College Press Service wire reporter: telephone interview.
Cunningham, Walter (2012, February 14). NASA Astronaut (Apollo 7): telephone interview.
Dotto, Lydia (2012, January 20), Toronto Globe and Mail Science Reporter: telephone interview.
Dunne, Dick (2013, May 2), Grumman PR director, 1967–1972: telephone interview.
Fine, Sam (2012, February 28), Bendix Corporation Communications: telephone interview.
Gentry, Michael (2011, November 21), NASA Public Affairs: telephone interview.
Harrison, Wayne (2012, January 19), KMHT Radio: telephone interview.
Jackson, Al (2012, October 10), Astrophysicist and NASA MSC employee: telephone interview.
Johnson, Elwood, (2012, June 11 & 13). Spacemobile program instructor, NASA contractor, and Fifty-State Apollo 11 tour participant.
King, Jack (2011, November 9), NASA Public Affairs: telephone interview.
Kirk, Preston (2012 August 30), UPI Houston Bureau reporter: telephone interview.
Larson, William (2012 January 9), ABC Radio News, Florida: telephone interview.

Thompson, Jeffrey (2012, January 25), KMSC Radio, Houston: telephone interview.
Turner, Thomas (2013, October 7), Past President, National Space Club, Consultant to Stanley Kubrick and Arthur C. Clarke on *2001: A Space Odyssey*.
Ward, Douglas (2011, November 29 & December 12), NASA Public Affairs: telephone interview.

JOHN F. KENNEDY LIBRARY ORAL HISTORY PROJECT

Glenn, John H. (1964, June 12), NASA Astronaut.
Shepard, Alan B. (1964, June 12), NASA Astronaut.

JOHNSON SPACE CENTER ORAL HISTORY INTERVIEWS

Anders, William A. (1997, October 8), NASA Astronaut.
Armstrong, Neil A. (2001, September 19), NASA Astronaut.
Barbree, Jay (2002, June 14), Journalist.
Barnes, Geneva B. (1999, March 26), NASA Public Affairs.
Bean, Alan L. (1998, June 23, & 2010, February 23), NASA Astronaut.
Biggs, Charles (2002, August 1), NASA Public Affairs.
Borman, Frank F. (1999, April 13), NASA Astronaut.
Carpenter, Scott M. (1998, March 30, & 1999, January 27), NASA Astronaut.
Cernan, Eugene A. (2007, December 11), NASA Astronaut.
Collins, Michael (1997, October 8), NASA Astronaut.
Cooper, Gordon L. (1998, May 21). NASA Astronaut.
Cunningham, R. Walter (1999, May 24), NASA Astronaut.
Engle, Joe H. (2004, April 22, May 5, May 27, June 3, June 24), NASA Astronaut.
Fruland, Walter S. (2009, September 24) NASA Public Affairs.
Glenn, John H. (1997, August 25), NASA Astronaut.
Gordon, Richard F. (1997, October 17, & 1999, June 16), NASA Astronaut.
Haise, Fred W. (1999, March 23), NASA Astronaut.
Haney, Paul. (2003, January 3), NASA Public Affairs.
Kraft, Christopher C. (2008, May 23), NASA Flight Crew Operations Chief.
Kranz, Eugene F. (1998, March 19, & 1999, January 8, April 28), NASA Apollo Flight Director.
Lovell, James A. (1999, May 25), NASA Astronaut.
Mattingly, Thomas K. (2001, January 6, & 2002, April 22), NASA Astronaut.
McCall, Robert T. (2000, March 28), NASA Space Artist Program participant.
McDivitt, James A. (1999, June 29), NASA Astronaut & Apollo Spacecraft Program Manager.
McLeaish, John E. (2001, November 15), NASA Public Affairs.
Mitchell, Edgar D. (1997, September 3), NASA Astronaut.
O'Hara, Dee. (2002, April 23), NASA Astronaut Nurse.
Parker, Louis A. (2011, December 6), NASA Public Affairs.
Riley, John E. (2002, October 9), NASA Public Information Specialist.
Schirra, Walter M. (1998, December 1), NASA Astronaut.
Schmitt, Harrison H. (1999, July 14, & 2000, March 16), NASA Astronaut.
Schweickart, Russell L. (1999, October 19, & 2000, March 8), NASA Astronaut.
Shepard, Alan B. (1998, February 20). NASA Astronaut.
Slezak, Terry (2009, July 29), NASA Photographer.
Stafford, Thomas P. (1998, October 15), NASA Astronaut.
Underwood, Richard W. (2000, October 17), NASA Photographic Technology Lab.
Worden, Alfred M. (2000, May 26), NASA Astronaut.

NATIONAL AIR & SPACE MUSEUM ORAL HISTORY INTERVIEW

Duff, Brian (April 24, 26, & May 1, 1989), NASA Public Affairs Officer.

WEBSITES

Apollo Lunar Surface Jounal. http://www.hq.nasa.gov/alsj/frame.html
Boeing Company, Apollo 11 recollections site. http://www.boeing.com/defense-space/space/apollo11/recollections.html
collectSpace: specifically the message boards and news center. http://www.collectspace.com
Heritage Archive: World's largest online newspaper archive (1753-present). http://newspaperarchive.com
Johnson Space Center News Roundup Archives – Manned Space Craft Center, Houston. (1961-2001). http://www.jsc.nasa.gov/history/roundups/roundups.htm
Johnson Space Center Mission Transcript Archives—Manned Space Craft Center, Houston. (Mercury – Apollo 17). http://www.jsc.nasa.gov/history/mission_trans/mission_transcripts.htm
National Archives Online Public Access Search. http://www.archives.gov/research/search/
New York Times online archive (1851-present). http://www.nytimes.com/ref/membercenter/nytarchive.html
YouTube: vintage astronaut and contractor commercials and NASA press footage. http://www.youtube.com

www.marketingthemoon.com
We invite our readers to visit our website.

Acknowledgments

Our deep gratitude for many years of camaraderie and fellowship on all things space, marketing and otherwise, goes to Leslie Cantwell, Leon Ford, Christopher Orwoll, Jason Rubin—and especially to Larry McGlynn, who allowed us access to his archives and who provided invaluable help throughout the project. We treasure your friendship.

We are especially grateful to our friends in the space artifact collecting community, particularly Robert Pearlman at collectSpace, Kim and Sally Poor at Astronaut Central and Spacefest, Steve Hankow of Farthest Reaches, Florian Noeller of Space Flori, Michael Constantine of Moonpans, Donnis Willis of Lunar Legacies, and Scott Schneeweis of Spaceaholic, and all the fine and dedicated people at the Astronaut Scholarship Foundation. We'd like to thank the many hard-working people at RR Auction, Heritage Auctions, Regency-Superior Galleries, and Bonham's for their many excellent auctions over the years, from which we have acquired much of our historic material—when not directly from the NASA and industry participants themselves.

We would love to thank individually all of our friends on the online space collector community collectSpace.com, but you are a cast of thousands and we dread offending even one of you. We salute the extraordinary dedication to space history that you exemplify and inspire. You are a global force—and an addiction!

Much of what we know has come from astronauts, former contractor PR and marketing professionals, journalists, and current and former NASA Public Affairs officers, who so graciously gave us their time, memories, and passion. They include not only the many who submitted to formal interviews specifically for *Marketing the Moon*, and who are listed in the bibliography, but also those with whom we have held numerous and detailed conversations over the years: Buzz Aldrin, Vance Brand, Jerry Carr, Charlie Duke, Ed Fendell, Owen Garriott, Ed Gibson, Dick Gordon, Gerry Griffin, Fred Haise, Joe Kerwin, Gene Kranz, Sy Libergott, Jack Lousma, Jim Lovell, Ed Mitchell, Dee O'Hara, Bill Pogue, Dave Scott, Tom Stafford, Paul Weitz, and Al Worden, as well as the late Neil Armstrong, Sam Beddingfield, and Guenter Wendt. Your generous conversations, whether over dinner or at a cocktail reception, have informed not only our approach to the material, but also our respect and admiration for the contributions of the hundreds of thousands of NASA employees and contractors with whom you worked.

Special shout outs go to Mike Gentry of NASA's photographic division, and Ed Hengeveld, a former journalist from The Netherlands, for their amazing photo research skills and access to their archives. Thanks to Betsy Sarles for her help when we were stuck, to Keith Jennings for his kind support, and to Seth Godin who inspired us to create art.

Well into the eleventh hour, Andrew Chaikin, one of the most experienced and distinguished writers about space, read the page proofs and made some excellent suggestions, which we incorporated with the utmost gratitude. Seldom does life get better than that!

We would also like to thank Captain Eugene A. Cernan, not only for participating in our formal interviews and engaging us over the years in many hours of conversation at dinners and events, but also for honoring us with a wise and inspiring foreword for our book.

At MIT Press, we'd like to thank Roger Conover, the executive editor who acquired and guided the book, and Katie Hope and Colleen Lanick who marketed and promoted it with knowledge, care, and sincere appreciation.

An extra-special thank you goes to Scott-Martin Kosofsky for his expertise and wisdom. Scott guided us through the entire publishing process in this, our first illustrated book. He acted as our agent and *über*editor, and created and executed the design that you see here, taking more than three thousands documents and images and curating them into a coherent narrative. Scott's longtime colleague Alan Andres was our literary and research editor, reading every word and checking every fact, making significant contributions to every aspect of the book. Alan loves history, and his extraordinary grasp of the Apollo era made this book far better than it would have been otherwise. Thank you, Scott and Alan—we couldn't have done it without you.

Finally, our thanks and love go to our families—Yukari, Allison, Karin, Tanja, and Philip—who make this amazing journey aboard spaceship Earth together all the more valuable and meaningful. This book is dedicated to you.